Donald Greenspan

Particle Modeling

1997

Birkhäuser

Boston • Basel • Berlin

Donald Greenspan
Department of Mathematics
University of Texas at Arlington
Arlington, Texas 76019

Library of Congress Cataloging-in-Publication Data

Greenspan, Donald.
    Particle modeling / Donald Greenspan.
       p. cm. -- (Modeling and simulation in science, engineering &
technology)
       Includes bibliographical references and index.
       ISBN 0-8176-3985-3 (hardcover : alk. paper). -- ISBN 3-7643-3985-3
       1. Science--Computer simulation. 2. Technology--Computer
simulation. 3. Differential equations--Numerical solutions.
4. Initial value problems--Numerical solutions. I. Title.
II. Series.
Q183.9.G74 1997
531'11'0113--DC21   97-20692
                                                      CIP

Printed on acid-free paper
© 1997 Birkhäuser Boston

Birkhäuser

ISBN 0-8176-3985-3
ISBN 3-7643-3985-3
Typeset by The Bartlett Press, Inc., Marietta, GA
Printed and bound by Hamilton Printing Co., Rensselaer, NY
Printed in the U.S.A.

9 8 7 6 5 4 3 2 1

CONTENTS

Preface

Contemporary science teaches that:

    (1) All things change with time.

    (2) All material bodies consist of atoms and/or molecules.

This book is concerned with computer simulation of scientific and engineering phenomena in a fashion which is consistent with principles (1) and (2), above. Our approach demands the approximate solution of initial value problems for systems of ordinary differential equations. The computers used for the examples to be discussed are the Digital Alpha275 personal computer and the Cray YMP/8.

The presentation is divided into three parts. The first part is concerned with mathematical, physical and numerical considerations and serves as a basis for the remainder of the book. The second part is concerned with the development of intuition, which is accomplished through extensive qualitative simulations and analyses. The third part is concerned with quantitative simulation of basic scientific and engineering phenomena. The penultimate chapter extends the approach to Special Relativity, but in a fashion which does not require previous study of the subject.

In general, Chapters 3-16 are independent of each other, so that the reader can study an individual application of interest without studying the other chapters of the presentation.

The approach developed here is distinctly different from that of classical continuum mechanics. Simulation is founded on discrete concepts only and is entirely consistent with modern theories of dynamical behavior.

Finally, I wish to thank the World Scientific Publishing Company, Singapore, for allowing me to use freely in this book related materials from my earlier book *Quasimolecular Modeling* (1991).

Donald Greenspan
Arlington, Texas
1997

Part I

Mathematical, Physical and Numerical Considerations

1

Particle Modeling:
What It Is and What It Is Not

1.1 Introduction

Our concern in this book is with a new area of simulation called *particle simulation* or *particle modeling* or even *discrete modeling*. Though specifics will follow later, we observe now, for the purpose of providing an overview, that particle modeling is the study of the dynamical behavior of solids and fluids in response to external forces, the solids and fluids being modeled as systems of atoms, or molecules, or aggregates of atoms and molecules. The dynamical equations are systems of second order, nonlinear, ordinary (rather that partial) differential equations.

Note that our usage of the term **particle** is different from the usage of others. Buneman et al. (1980) and Hockney and Eastwood (1981) use the term **particle** to represent an ion in a plasma. Amsden (1966) and Harlow and Sanmann (1965) use the term to represent a fluid point of positive mass which moves in accordance with mass, energy and momentum conservation properties which are incorporated in a system of partial differential equations in two space dimensions.

1.2 Classical Molecular Forces

From the classical, Newtonian point of view, both atoms and molecules exhibit the following behavior. Two molecules, for example, interact only locally, that is, when they are in close proximity to each other. Qualitatively, this interaction is of the following character (Feynman, Leighton and Sands (1963)). If pushed together, the molecules repel; if pulled apart, they attract; and the repulsive force is of a greater order of magnitude than is the attractive one. A mathematical formulation of this behavior can be given as follows (Hirschfelder, Curtiss and Bird (1965)).

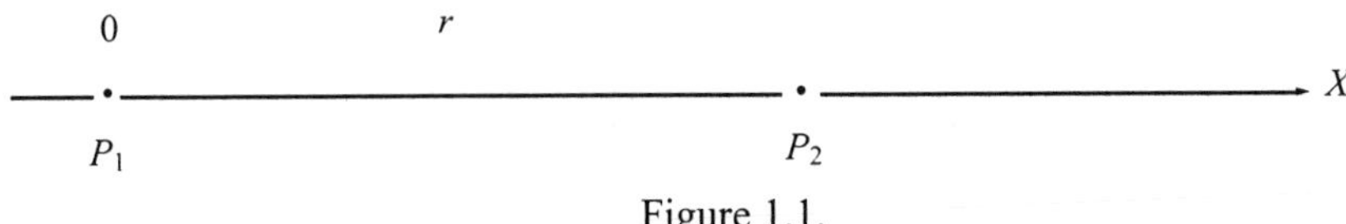

Figure 1.1.

Consider two molecules P_1 and P_2 on an X-axis, as shown in Fig. 1.1. Let P_1 be at the origin and let P_2 be at a positive distance r from P_1. Let the force $\vec{F}$ which P_1 exerts on P_2 have magnitude F given by

$$F = -\frac{G}{r^p} + \frac{H}{r^q}, \tag{1.1}$$

where G, H, p, q are positive constants with $q > p$. Consider, for example, $G = H = 1$, $p = 7$, $q = 13$, which are good approximations for a variety of experimental results (Hirschfelder, Curtiss and Bird (1965)). Then

$$F = -\frac{1}{r^7} + \frac{1}{r^{13}}. \tag{1.2}$$

If, in (1.2), $r = 1$, then $F = 0$, so that P_1 exerts no force on P_2. In this case, one says that the molecules are in equilibrium. If $r > 1$, say $r = 2$, then

$$F = -\frac{1}{2^7} + \frac{1}{2^{13}}, \tag{1.3}$$

which is negative, so that P_1 exerts an attractive force on P_2. If, on the other hand, $0 < r < 1$, say, $r = 0.1$, then

$$F = -\frac{1}{(0.1)^7} + \frac{1}{(0.1)^{13}}, \tag{1.4}$$

which is positive, so that P_1 exerts a repulsive force on P_2. As r approaches zero, the force F in (1.2) becomes unbounded in magnitude. Mathematically, r is not allowed to be zero because, if it were, F in (1.2) would be undefined. Physically, r is not allowed to be zero because one assumes conservation of mass, so that the same position cannot be occupied simultaneously by different physical entities. If one sets $F = 0$ in (1.1), then, using the reasoning above for (1.2), one finds that equilibrium results if

$$r = \left(\frac{H}{G}\right)^{\frac{1}{(q-p)}}, \tag{1.5}$$

with an attractive force resulting for larger values of r and a repulsive force for the smaller ones.

It is important to observe that even though the gross motion of, for example, a fluid may be physically stable, the motion between two neighboring molecules of the fluid, in accordance with (1.1), may be highly volatile. This volatility, however, is strictly local.

In general, and for consistency, we will employ *cgs* units throughout. Thus, let P_1, P_2 be two particles, r *cm* apart, in three-dimensional xyz space. To P_1 and P_2 let there be associated a potential $\phi(r)$, which depends only on r. Let the units of ϕ be *ergs* ($= \frac{gcm^2}{sec^2}$). Then the force $\vec{F}$ between P_1 and P_2 will be given in dynes ($= \frac{gcm}{sec^2}$) and the magnitude F of $\vec{F}$ satisfies

$$F = -\frac{d\phi}{dr}$$

1.3 General Modeling Principles

To simulate the dynamical response of a material body, that is, a solid or fluid, to external forces, we will proceed at first as follows. Assume that the body consists of N particles. Denote these by P_i, $i = 1, 2, \ldots, N$ and let the mass of P_i be m_i. From given initial data, the motion of each P_i is then prescribed by the coupled system of ordinary differential equations:

$$\vec{F}_i = m\ddot{\vec{r}}_i \quad i = 1, 2, \ldots, N, \tag{1.6}$$

in which $\vec{F}_i$ is the force on P_i, $\vec{r}_i$ is the position vector of P_i, and differentation is with respect to time. In (1.6), we assume that

$$\vec{F}_i = \vec{F}_i^{**} + \vec{F}_i^{*} \tag{1.7}$$

where $\vec{F}_i^{**}$ is an external or long range force, which, like gravity, can act on all the particles uniformly or, like a driving force, can act on a particular subset of the particles; and $\vec{F}_i^{*}$ is a local, short range force on P_i due to interaction with its immediate neighbors. In practice, a positive parameter D, called the **distance of local interaction parameter**, will often be associated with $\vec{F}_i^{*}$. It will assure that $\vec{F}_i^{*}$ is, in fact, local, by only allowing particles whose distance to P_i is less than D to have a nonzero effect on P_i. Hence, D can be viewed as a switching parameter which turns off $\vec{F}_i^{*}$ for all particles except those close to P_i.

2

Numerical Methodology

2.1 Introduction

It will be necessary in particle modeling to solve the system of second order differential equations (1.6) from given initial data. Since, in general, $\vec{F}$ will be nonlinear, numerical methodology will be essential. Our choice of numerical methods will be guided by the following two observations. First, since ϕ depends only on r, the system (1.6) is completely conservative (Goldstein (1980)), that is, the system's energy, linear momentum, and angular momentum are time invariants. Second, for many potentials and forces to be considered, $r = 0$ will be a singularity in the sense that the potential or the magnitude of the force becomes unbounded as r goes to zero. For such cases the time step must be small for values of r close to zero, which precludes the basic value of high order numerical techniques. For these reasons we will employ only the two methods to be discussed in this chapter, that is, the *leap frog method*, which is basically a central difference, low order method which is highly efficient and easy to program, and a *completely conservative method* which conserves exactly the same invariants as (1.6).

Throughout, let $h = \Delta t$ be a positive time step. Let $t_k = k\Delta t$, $k = 0, 1, 2, \ldots$. For $i = 1, 2, \ldots N$, let P_i have mass m_i and at t_k, let P_i be located at $\vec{r}_{i,k}$, have velocity $\vec{v}_{i,k}$ and have acceleration $\vec{a}_{i,k}$. If the vector from P_i to P_j at t_k is denoted by $\vec{r}_{ij,k}$, we define its magnitude by $r_{ij,k} = \| \vec{r}_{ij,k} \|$.

Let us turn first to the leap frog method.

2.2 The Leap Frog Method

The leap frog formulas, which relate position, velocity and acceleration for $i = 1, 2, \ldots, N$ are (Greenspan (1980a)):

$$\vec{v}_{i,\frac{1}{2}} = \vec{v}_{i,0} + \frac{(\Delta t)}{2}\vec{a}_{i,0} \qquad \text{(starter formula)} \tag{2.1}$$

$$\vec{v}_{i,k+1/2} = \vec{v}_{i,k-1/2} + (\Delta t)\vec{a}_{i,k}, \qquad k = 1, 2, 3, \ldots \tag{2.2}$$

$$\vec{r}_{i,k+1} = \vec{r}_{i,k} + (\Delta t)\vec{v}_{i,k+1/2}, \qquad k = 0, 1, 2, \ldots. \tag{2.3}$$

The name "leap frog" derives from the way position and velocity are defined at alternate, sequential time points. Note also that if (2.2) is solved for $\vec{a}_{i,k}$ and (2.3) is solved for $\vec{v}_{i,k+1/2}$, then the resulting formulas are central difference, $O((\Delta t)^2)$ approximation formulas.

If at time t_k one rewrites (1.6) and (1.7), respectively, as

$$\vec{F}_{i,k} = m_i\vec{a}_{i,k}, \qquad i = 1, 2, \ldots, N \tag{2.4}$$

$$\vec{F}_{i,k} = \vec{F}_{i,k}^{**} + \vec{F}_{i,k}^{*}, \qquad i = 1, 2, \ldots, N \tag{2.5}$$

then (2.1)–(2.5) determine the positions and velocities of all N particles recursively and explicitly from given initial data.

Example 2.1

To illustrate the numerical procedure to be followed, consider the following simple example in only one space dimension. On an X-axis, let P_1 and P_2, with masses $m_1 = 2$, $m_2 = 1$, be located initially at $x_{1,0} = 0$, $x_{2,0} = 1$ and have initial velocities $v_{1,0} = -1$, $v_{2,0} = 3$. Let the distance of local interaction be $D = 1.5$ and set $\Delta t = 0.1$. Let the forces on P_1 and P_2 at t_k be given by

$$F_{1,k}^{**} = -980, \tag{2.6}$$

$$F_{1,k}^{*} = \left(-\frac{1}{|x_{2,k} - x_{1,k}|^3} + \frac{2}{|x_{2,k} - x_{1,k}|^6}\right)\frac{x_{1,k} - x_{2,k}}{|x_{2,k} - x_{1,k}|}, \tag{2.7}$$

$$F_{2,k}^{**} = 0, \tag{2.8}$$

$$F_{2,k}^{*} = -F_{1,k}^{*}. \tag{2.9}$$

Then, from (2.1)–(2.4), one finds for P_1 that

$$v_{1,\frac{1}{2}} = v_{1,0} + (0.05)a_{1,0}, \qquad \text{(Starter formula)}$$

$$v_{1,k+1/2} = v_{1,k-1/2} + (0.1)a_{1,k}, \qquad k = 1, 2, 3, \ldots$$

$$x_{1,k+1} = x_{1,k} + (0.1)v_{1,k+1/2}, \qquad k = 0, 1, 2, \ldots$$

or, equivalently, that

$$v_{1,1/2} = -1 + (0.05)(F_{1,0}/2) = -1 + (0.025)F_{1,0}, \tag{2.10}$$

$$v_{1,k+1/2} = v_{1,k-1/2} + (0.1)(F_{1,k}/2) = v_{1,k-1/2} + (0.05)F_{1,k}, \tag{2.11}$$

$$x_{1,k+1} = x_{1,k} + (0.1)v_{1,k+1/2}. \tag{2.12}$$

Since $|x_{2,0} - x_{1,0}| < 1.5 = D$, it follows from (2.5)–(2.7) that

$$F_{1,0} = F_{1,0}^{**} + F_{1,0}^{*}$$

$$= -980 + \left(-\frac{1}{|x_{2,0} - x_{1,0}|^3} + \frac{2}{|x_{2,0} - x_{1,0}|^6} \right) \frac{x_{1,0} - x_{2,0}}{|x_{2,0} - x_{1,0}|}$$

$$= -980 + (1)(-1) = -981.$$

Thus, from (2.10)

$$v_{1,1/2} = -1 + (0.025)(-981) = -25.525. \tag{2.13}$$

One finds in an analogous fashion that

$$v_{2,1/2} = v_{2,0} + (0.05)a_{2,0} \,,$$

$$v_{2,k+1/2} = v_{2,k-1/2} + (0.1)a_{2,k} \,,$$

$$x_{2,k+1} = x_{2,k} + (0.1)v_{2,k+1/2} \,,$$

and

$$v_{2,1/2} = 3 + (0.05)(F_{2,0}/1) = 3 + (0.05)F_{2,0} \,, \tag{2.14}$$

$$v_{2,k+1/2} = v_{2,k-1/2} + (0.1)(F_{2,k}/1) = v_{2,k-1/2} + (0.1)F_{2,k} \,, \tag{2.15}$$

$$x_{2,k+1} = x_{2,k} + (0.1)v_{2,k+1/2} \,. \tag{2.16}$$

Since

$$F_{2,0} = F_{2,0}^{**} + F_{2,0}^{*} \,,$$

it follows from (2.6), (2.8) and (2.9) that

$$F_{2,0} = 0 - F_{1,0}^{*} = 1 \,.$$

Then, from (2.14)

$$v_{2,1/2} = 3.05 \,. \tag{2.17}$$

Thus, the velocities $v_{1,1/2}$ and $v_{2,1/2}$ of P_1 and P_2 at the time $t = 1/2$ have now been determined and are given by (2.13) and (2.17). The formulas (2.12) and (2.16) with $k = 0$ now yield the new positions $x_{1,1}$ and $x_{2,1}$ of P_1 and P_2, as follows:

$$x_{1,1} = x_{1,0} + (0.1)v_{1,1/2} = 0 + (0.1)(-25.525) = -2.5525, \qquad (2.18)$$

$$x_{2,1} = x_{2,0} + (0.1)v_{2,1/2} = 1 + (0.1)(3.05) = 1.305. \qquad (2.19)$$

The process now continues to determine next $v_{1,3/2}, v_{2,3/2}$. But since the distance $|x_{2,1} - x_{1,1}| = 3.8575 > 1.5 = D$, the switch is applied so that

$$F_{1,1}^* = F_{2,1}^* = 0 . \qquad (2.20)$$

Observe also that the notation in (2.20) should always remain clear if one remembers that the *first* subscript is always the *particle number* and the *second* is always the *time step*.

Once formulas (2.10) and (2.14) have been used to determine $v_{1,1/2}$ and $v_{2,1/2}$, they are no longer used. All the remaining trajectory calculations are done with (2.11), (2.12), (2.15), and (2.16). Hence, the counter is now set to $k = 1$. From (2.11) and (2.15), then,

$$v_{1,3/2} = v_{1,1/2} + (0.05)F_{1,1}^{**} = -25.525 + (0.05)(-980) = -74.525, \quad (2.21)$$

$$v_{2,3/2} = v_{2,1/2} - (0.01)F_{2,1}^{**} = 3.05 - (0.01)(0) = 3.05 . \qquad (2.22)$$

Now, having the velocities of P_1 and P_2 at $t = 3/2$, we find their new positions from (2.12) and (2.16) to be

$$x_{1,2} = x_{1,1} + (0.1)v_{1,3/2} = -2.5525 + (0.1)(-74.525) = -10.005 , \quad (2.23)$$

$$x_{2,2} = x_{2,1} + (0.1)v_{2,3/2} = 1.305 + (0.1)(3.05) = 1.610. \qquad (2.24)$$

The counter is then increased to $k = 2$ and the iteration continues in the indicated fashion.

With regard to the leap frog formulas and their application, several relevant observations must now be made. First, note that (2.1)–(2.3) have been given in vector form, so that they can be applied in 1, 2 or 3 space dimensions, as needed. Of course, in two dimensions, one would have

$$\vec{r}_{i,k} = (x_{i,k}, y_{i,k})$$

$$\vec{v}_{i,k} = (v_{i,x,k}, v_{i,y,k})$$

$$\vec{a}_{i,k} = (a_{i,x,k}, a_{i,y,k})$$

$$\vec{F}_{i,k} = (F_{i,x,k}, F_{i,y,k})$$

while in three dimensions one need only append a z-component to the above formulas.

Note also that for relatively large N, the determination of the nearest neighbor for each particle of a system will usually be the most time consuming part of any simulation. This is particularly valid in simulations of fluids. Indeed, when one simulates a solid, the near neighbors of any P_1 can be given uniquely and explicity for all time. But when one simulates a fluid, this is not the case. For this reason, there have been a variety of "economical near-neighbor" algorithms developed recently for simulations of fluids (e.g., Boris (1986)). However, in each case, one either does not include all the neighbors, or one is forced to alter the particle ordering, or one does not determine the near neighbors at every time step. In the case when one alters the partial ordering, for example, one cannot follow the trajectory of any particular particle from an initial to a later time, and this capability is desirable, since we may wish to explore the motion of individual particles at the onset of turbulence. Thus, we will not take advantage of existing "near-neighbor" algorithms.

In Appendices A1 and A2 two typical FORTRAN programs which use the leap frog formula are given. These programs are useful in Sect. II.

2.3 Completely Conservative Numerical Methodology

For clarity, we proceed in three dimensions with a fundamental N-body problem, that is, with $N = 3$. Extension to arbitrary N follows using entirely similar ideas and proofs as for $N = 3$.

For $i = 1, 2, 3$, let P_i of mass m_i be at $\vec{r}_i = (x_i, y_i, z_i)$ at time t. Let the positive distance between P_i and P_j, $i \neq j$, be r_{ij}, with $r_{ij} = r_{ji}$. Let $\phi(r_{ij}) - \phi_{ij}$, given in ergs, be a potential for the pair P_i, P_j. Then the Newtonian dynamical equations for the three-body interactions are the three vector (nine scalar) second order differential equations

$$m_i \frac{d^2\vec{r}_i}{dt^2} = -\frac{\partial\phi}{\partial r_{ij}} \frac{\vec{r}_i - \vec{r}_j}{r_{ij}} - \frac{\partial\phi}{\partial r_{ik}} \frac{\vec{r}_i - \vec{r}_k}{r_{ik}}, \quad i = 1, 2, 3, \qquad (2.25)$$

where $j = 2$ and $k = 3$ when $i = 1$; $j = 1$ and $k = 3$ when $i = 2$; $j = 1$ and $k = 2$ when $i = 3$.

System (2.25) conserves energy, linear momentum, and angular momentum. In addition, it is covariant, that is, it has the same functional form under translation, rotation, and uniform relative motion of coordinate frames. Our problem is to devise a numerical scheme for solving system (2.25) from given initial data so that the numerical scheme preserves the very same system invariants. We will also show later that the numerical method developed is also covariant.

For $h > 0$, let $t_n = nh$, $n = 0, 1, 2, \ldots$. At time t_n, let P_i be at $\vec{r}_{i,n} = (x_{i,n}, y_{i,n}, z_{i,n})$ and have velocity $\vec{v}_{i,n} = (v_{i,x,n}, v_{i,y,n}, v_{i,z,n})$. Let distances $|P_1 P_2|, |P_1 P_3|, |P_2 P_3|$ be denoted by $r_{12,n}, r_{13,n}, r_{23,n}$, respectively. We now approximate the second order differential system (2.25) by the first order difference system

$$\frac{\vec{r}_{i,n+1} - \vec{r}_{i,n}}{\Delta t} = \frac{\vec{v}_{i,n+1} + \vec{v}_{i,n}}{2} \tag{2.26}$$

$$m_i \frac{\vec{v}_{i,n+1} - \vec{v}_{i,n}}{\Delta t} = -\frac{\phi(r_{ij,n+1}) - \phi(r_{ij,n})}{r_{ij,n+1} - r_{ij,n}} \frac{\vec{r}_{i,n+1} + \vec{r}_{i,n} - \vec{r}_{j,n+1} - \vec{r}_{j,n}}{r_{ij,n+1} + r_{ij,n}} \tag{2.27}$$

$$-\frac{\phi(r_{ik,n+1}) - \phi(r_{ik,n})}{r_{ik,n+1} - r_{ik,n}} \frac{\vec{r}_{i,n+1} + \vec{r}_{i,n} - \vec{r}_{k,n+1} - \vec{r}_{k,n}}{r_{ik,n+1} + r_{ik,n}},$$

with $j = 2$ and $k = 3$ when $i = 1$; $j = 1$ and $k = 3$ when $i = 2$; $j = 1$ and $k = 2$ when $i = 3$.

System (2.26) and (2.27) constitutes 18 implicit recursion equations for the unknowns $x_{i,n+1}$, $y_{i,n+1}$, $z_{i,n+1}$, $v_{i,x,n+1}$, $v_{i,y,n+1}$, $v_{i,z,n+1}$ in the 18 knowns $x_{i,n}$, $y_{i,n}$, $z_{i,n}$, $v_{i,x,n}$, $v_{i,y,n}$, $v_{i,z,n}$, $i = 1, 2, 3$. These equations can be solved readily by Newton's method (Greenspan (1980a)) to yield the numerical solution. (See also Appendices A3–A5.)

Let us show now that the numerical solution generated by (2.26) and (2.27) conserves the same energy, linear momentum, and angular momentum as does (2.25).

Consider first energy conservation. For this purpose, define

$$W_N = \sum_{n=0}^{N-1} \left\{ \sum_{i=1}^{3} m_i (\vec{r}_{i,n+1} - \vec{r}_{i,n}) \cdot (\vec{v}_{i,n+1} - \vec{v}_{i,n})/h \right\} . \tag{2.28}$$

Note, relative to (2.28), that since we are considering specifically the three-body problem, the symbol N in the summation (2.28) is now simply a numerical time index. Then, insertion of (2.26) into (2.28) yields

$$W_N = \frac{1}{2} m_1 v_{1,N}^2 + \frac{1}{2} m_2 v_{2,N}^2 + m_3 v_{3,N}^2 - \frac{1}{2} m_1 v_{1,0}^2 - \frac{1}{2} m_2 v_{2,0}^2 - \frac{1}{2} m_3 v_{3,0}^2 ,$$

so that

$$W_N = K_N - K_0 . \tag{2.29}$$

Insertion of (2.27) into (2.28) implies, with some tedious but elementary algebraic manipulation,

$$W_N = \sum_{n=0}^{N-1} (-\phi_{12,n+1} - \phi_{13,n+1} - \phi_{23,n+1} + \phi_{12,n} + \phi_{13,n} + \phi_{23,n}) ,$$

so that

$$W_N = -\phi_N + \phi_0 . \tag{2.30}$$

Elimination of W_N between (2.29) and (2.30) then yields conservation of energy, that is,

$$K_N + \phi_N = K_0 + \phi_0, \quad N = 1, 2, 3, \ldots .$$

Moreover, since K_0 and ϕ_0 only depend on initial data, it follows that K_0 and ϕ_0 are the same in both the continuous and the discrete cases, so that the energy conserved by the numerical method is exactly that of the continuous system. Note also that the proof was independent of h. Thus we have proved the following theorem.

THEOREM 2.1
Independently of h, (2.26) and (2.27) are energy conserving, that is

$$K_N + \phi_N = K_0 + \phi_0, \quad N = 1, 2, \ldots.$$

Next, the linear momentum $\vec{M}_i(t_n) = \vec{M}_{i,n}$ of P_i at t_n is defined to be the vector

$$\vec{M}_{i,n} = m_i(v_{i,x,n}, v_{i,y,n}, v_{i,z,n}) . \tag{2.31}$$

The linear momentum $\vec{M}_n$ of the three-body system at time t_n is defined to be the vector

$$\vec{M}_n = \sum_{i=1}^{3} \vec{M}_{i,n} . \tag{2.32}$$

Now, from (2.27),

$$m_1(\vec{v}_{1,n+1} - \vec{v}_{1,n}) + m_2(\vec{v}_{2,n+1} - \vec{v}_{2,n}) + m_3(\vec{v}_{3,n+1} - \vec{v}_{3,n}) \equiv \vec{0} . \tag{2.33}$$

Thus, in particular, for $n = 0, 1, 2, \ldots,$

$$m_1(v_{1,x,n+1} - v_{1,x,n}) + m_2(v_{2,x,n+1} - v_{2,x,n}) + m_3(v_{3,x,n+1} - v_{3,x,n}) = 0 . \tag{2.34}$$

Summing both sides of (2.34) from $n = 0$ to $n = N - 1$ implies

$$m_1 v_{1,x,N} + m_2 v_{2,x,N} + m_3 v_{3,x,N} = C_1 , \quad N \geq 1, \tag{2.35}$$

in which

$$m_1 v_{1,x,0} + m_2 v_{2,x,0} + m_3 v_{3,x,0} = C_1 . \tag{2.36}$$

Similarly,

$$m_1 v_{1,y,N} + m_2 v_{2,y,N} + m_3 v_{3,y,N} = C_2 . \tag{2.37}$$

$$m_1 v_{1,z,N} + m_2 v_{2,z,N} + m_3 v_{3,z,N} = C_3 , \tag{2.38}$$

in which

$$m_1 v_{1,y,0} + m_2 v_{2,y,0} + m_3 v_{3,y,0} = C_2 . \tag{2.39}$$

$$m_1 v_{1,z,0} + m_2 v_{2,z,0} + m_3 v_{3,z,0} = C_3 . \tag{2.40}$$

Thus,

$$\vec{M}_n = \sum_{i=1}^{3} \vec{M}_{i,n} = (C_1, C_2, C_3) = \vec{M}_0 , \quad n = 1, 2, 3, \ldots,$$

which is the classical law of conservation of linear momentum. Note again that M_0 depends only on the initial data.

Thus, we have the following theorem.

THEOREM 2.2
Independently of h, (2.26) and (2.27) conserve linear momentum, that is,

$$\vec{M}_n = \vec{M}_0 , \quad n = 1, 2, 3, \ldots .$$

We turn finally to angular momentum. The angular momentum $\vec{L}_{i,n}$ of P_i at t_n is defined to be the vector

$$\vec{L}_{i,n} = m_i (\vec{r}_{i,n} \times \vec{v}_{i,n}) . \tag{2.41}$$

The angular momentum of a three-body system at t_n is defined to be the vector

$$\vec{L}_n = \sum_{i=1}^{3} \vec{L}_{i,n}. \tag{2.42}$$

It then follows readily (Greenspan (1980a)) that

$$\vec{L}_{i,n+1} - \vec{L}_{i,n} = \frac{1}{2} m_i (\vec{r}_{i,n+1} + \vec{r}_{i,n}) \times (\vec{v}_{i,n+1} - \vec{v}_{i,n}) . \tag{2.43}$$

Thus,

$$\vec{L}_{n+1} - \vec{L}_n = \frac{1}{2} \sum_{i=1}^{3} m_i (\vec{r}_{i,n+1} + \vec{r}_{i,n}) \times (\vec{v}_{i,n+1} - \vec{v}_{i,n}) . \tag{2.44}$$

However, substitution of (2.27) into (2.44) yields, after some tedious calculation,

$$\vec{L}_{n+1} - \vec{L}_n = \vec{0} , \quad n = 1, 2, \ldots ,$$

which implies, independently of h, the conservation of angular momentum. Note again that $\vec{L}_0$ depends only on initial data. Thus the following theorem has been proved.

THEOREM 2.3
Independently of h, (2.26) and (2.27) conserve angular momentum, that is,

$$\vec{L}_n = \vec{L}_0 , \quad n = 1, 2, 3, \ldots .$$

Let us turn now to covariance and begin the discussion as simply as possible. When a dynamical equation is structurally invariant under a transformation, the equation is said to be *covariant* or *symmetric*. The transformations we will consider are translation, rotation and uniform relative motion. We will concentrate only on two-dimensional systems, because the related techniques and results extend directly to three dimensions. A *general* Newtonian force will be considered, so that the particular type of force discussed thus far will be included in the discussion as a special case. Finally, we will concentrate only on the motion of a single particle P of mass m, with the extension to the N-body problem following in a natural way. And though the assumptions just made may seem to be excessive, it will be seen shortly that they make the required mathematical methodology readily transparent.

Suppose now that a particle P of mass m is in motion in the XY plane and that for $\Delta t = h > 0$ its motion from given initial data is determined by a force $\vec{F}(t_n) = \vec{F}_n = (F_{x,n}, F_{y,n})$ and by the dynamical difference equations

$$F_{x,n} = m(v_{x,n+1} - v_{x,n})/h \tag{2.45}$$

$$F_{y,n} = m(v_{y,n+1} - v_{y,n})/h \ . \tag{2.46}$$

Our problem is as follows. Let $x = f_1(x^*, y^*)$, $y = f_2(x^*, y^*)$ be a change of coordinates. Under this transformation, let $F_{x,n} = F^*_{x^*,n}$, $F_{y,n} = F^*_{y^*,n}$. Then we will want to prove that in the X^*Y^* system, the dynamical equations of motion are

$$F^*_{x^*,n} = m(v_{x^*,n+1} - v_{x^*,n})/h \tag{2.47}$$

$$F^*_{y^*,n} = m(v_{y^*,n+1} - v_{y^*,n})/h \ , \tag{2.48}$$

which will establish covariance.

In consistency with (2.26), we have, in particular

$$\frac{x_{n+1} - x_n}{h} = \frac{v_{x,n+1} + v_{x,n}}{2}, \quad \frac{x^*_{n+1} - x^*_n}{h} = \frac{v_{x^*,n+1} + v_{x^*,n}}{2}, \tag{2.49}$$

$$\frac{y_{n+1} - y_n}{h} = \frac{v_{y,n+1} + v_{y,n}}{2}, \quad \frac{y^*_{n+1} - y^*_n}{h} = \frac{v_{y^*,n+1} + v_{y^*,n}}{2}. \tag{2.50}$$

Relative to (2.49) and (2.50), the following lemma will be of value.

LEMMA 2.1
Equations (2.49) and (2.50) imply

$$v_{x,1} = \frac{2}{h}(x_1 - x_0) - v_{x,0}; \ v_{x^*,1} = \frac{2}{h}(x^*_1 - x^*_0) - v_{x^*,0} \tag{2.51}$$

$$v_{y,1} = \frac{2}{h}(y_1 - y_0) - v_{y,0}; \ v_{y^*,1} = \frac{2}{h}(y^*_1 - y^*_0) - v_{y^*,0} \tag{2.52}$$

$$v_{x,n} = \frac{2}{h}\left[x_n + (-1)^n x_0 + 2\sum_{j=1}^{n-1}(-1)^j x_{n-j}\right] + (-1)^n v_{x,0}, \quad n \geq 2 \tag{2.53}$$

$$v_{x^*,n} = \frac{2}{h}\left[x^*_n + (-1)^n x^*_0 + 2\sum_{j=1}^{n-1}(-1)^j x^*_{n-j}\right] + (-1)^n v_{x^*,0}, \quad n \geq 2$$

$$v_{y,n} = \frac{2}{h}\left[y_n + (-1)^n y_0 + 2\sum_{j=1}^{n-1}(-1)^j y_{n-j} \right] + (-1)^n v_{y,0}, \quad n \geq 2 \quad (2.54)$$

$$v_{y^*,n} = \frac{2}{h}\left[y_n^* + (-1)^n y_0^* + 2\sum_{j=1}^{n-1}(-1)^j y_{n-j}^* \right] + (-1)^n v_{y^*,0}, \quad n \geq 2 \, .$$

The proof is immediate by (2.49), (2.50) and mathematical induction.

THEOREM 2.4
Equations (2.45) and (2.46) are covariant relative to the translation

$$x^* = x - a, \, y^* = y - b; \, a, b \text{ constants.} \tag{2.55}$$

PROOF   Define $v_{x,0} = v_{x^*,0}$, $v_{y,0} = v_{y^*,0}$. Then, from (2.51)

$$v_{x,1} = \frac{2}{h}[(x_1^* + a) - (x_0^* + a)] - v_{x^*,0}$$

$$= v_{x^*,1} \, .$$

Similarly,

$$v_{y,1} = v_{y^*,1} \, .$$

For $n \geq 2$, (2.53) yields

$$v_{x,n} = \frac{2}{h}\left[(x_n^* + a) + (-1)^n(x_0^* + a) + 2\sum_{j=1}^{n-1}(-1)^j(x_{n-j}^* + a) \right] + (-1)^n v_{x^*,0}.$$

$$\tag{2.56}$$

However, by the lemma, for both n even and n odd, (2.56) implies

$$v_{x,n} = v_{x^*,n} \, .$$

Similarly,

$$v_{y,n} = v_{y^*,n} \, .$$

Thus, for all $n = 0, 1, 2, 3, \ldots,$

$$v_{x,n} = v_{x^*,n} \tag{2.57}$$

$$v_{y,n} = v_{y^*,n}. \tag{2.58}$$

Thus,

$$F_{x^*,n}^* = F_{x,n} = m\frac{v_{x,n+1} - v_{x,n}}{h} = m\frac{v_{x^*,n+1} - v_{x^*,n}}{h} \, ,$$

Similarly,

$$F^*_{y^*,n} = m \frac{v_{y^*,n+1} - v_{y^*,n}}{h} \; ,$$

and the theorem is proved. $\square$

THEOREM 2.5
Under the rotation

$$x^* = x \cos \theta + y \sin \theta \tag{2.59}$$

$$y^* = y \cos \theta - x \sin \theta$$

where θ is the smallest positive angle measured in the counterclockwise direction from the X to the X^* axis, equations (2.45)–(2.46) are covariant.

PROOF   The proof follows along the same lines as that of Theorem 2.4 after one defines $v_{x^*,0}$ and $v_{y^*,0}$ by

$$v_{x^*,0} = v_{x,0} \cos \theta + v_{y,0} \sin \theta \tag{2.60}$$

$$v_{y^*,0} = v_{y,0} \cos \theta - v_{x,0} \sin \theta \; ,$$

and observes that

$$F^*_{x^*,n} = F_{x,n} \cos \theta + F_{y,n} \sin \theta \tag{2.61}$$

$$F^*_{y^*,n} = F_{y,n} \cos \theta - F_{x,n} \sin \theta. \quad \square$$

THEOREM 2.6
Under relative uniform motion of coordinate systems, equations (2.45) and (2.46) are covariant.

PROOF   Consider first motion in one dimension. Assume then that the X and X^* axes are in relative motion defined by

$$x_n^* = x_n - ct_n, \quad n = 0, 1, 2, \ldots, \tag{2.62}$$

in which c is a positive constant. If $v_{x,0}$ is the initial velocity of P along the X axis, let $v_{x^*,0}$ along the X^* axis be defined by

$$v_{x^*,0} = v_{x,0} - c \; . \tag{2.63}$$

Hence, for $n = 1$,

$$v_{x,1} = \frac{2}{h} \left[(x_1^* + ct_1) - (x_0^* + ct_0) \right] - v_{x,0} = v_{x^*,1} + c \; .$$

For $n \geq 2$,

$$v_{x,n} = \frac{2}{h} \left\{ x_n^* + (-1)^n x_0^* + 2 \sum_{j=1}^{n-1} (-1)^j x_{n-j}^* \right\} + (-1)^n v_{x,0}$$

$$+ \frac{2c}{h} \left\{ t_n + (-1)^n t_0 + 2 \sum_{j=1}^{n-1} (-1)^j t_{n-j} \right\} .$$

But,

$$t_n + (-1)^n t_0 + 2 \sum_{j=1}^{n-1} (-1)^j t_{n-j} = \begin{cases} 0, n & \text{even} \\ h, n & \text{odd.} \end{cases}$$

Thus, with the aid of Lemma 2.1, it follows that for both n odd and even

$$v_{x,n} = v_{x^*,n} + c .$$

Thus, for all $n = 0, 1, 2, 3, \ldots$,

$$F_{x^*,n}^* = F_{x,n} = m \frac{v_{x^*,n+1} + c - v_{x^*,n} - c}{h} = m \frac{v_{x^*,n+1} - v_{x^*,n}}{h} .$$

Under the assumption that

$$y^* = y - dt_n$$

in which d is a constant, one finds similarly that

$$F_{y^*,n}^* = m \frac{v_{y^*,n+1} - v_{y^*,n}}{h} ,$$

and the covariance is established. $\square$

Finally let us give a simple example to illustrate the methodology of this section. Consider the one-dimensional initial value problem for a particle P of unit mass in motion along the x-axis:

$$\ddot{x} = x^2 \tag{2.64}$$

$$x(0) = \dot{x}(0) = 1. \tag{2.65}$$

Then, $F(x) = x^2$. Choose $\phi(x) = -\frac{x^3}{3}$, so that

$$-\frac{\phi(x_{k+1}) - \phi(x_k)}{x_{k+1} - x_k} = \frac{1}{3} (x_{k+1}^2 + x_{k+1} x_k + x_k^2) . \tag{2.66}$$

The equations (2.26), (2.27) then reduce to

$$x_{k+1} - x_k - \frac{1}{2} h (v_{k+1} + v_k) = 0 \tag{2.67}$$

$$v_{k+1} - v_k - \frac{h}{3} (x_{k+1}^2 + x_{k+1} x_k + x_k^2) = 0 , \tag{2.68}$$

which are two equations for x_{k+1}, v_{k+1} in terms of x_k, v_k. Newtonian iteration formulas for this system are

$$x_{k+1}^{(n+1)} = x_k + \frac{1}{2}h(v_{k+1}^{(n)} + v_k)$$

$$v_{k+1}^{(n+1)} = v_k + \frac{h}{3}\left[(x_{k+1}^{(n+1)})^2 + (x_{k+1}^{(n+1)})(x_k) + (x_k)^2\right].$$

The initial guess for each iterative step is

$$x_{k+1}^{(0)} = x_k, \quad v_{k+1}^{(0)} = v_k .$$

For $h = 0.01$, the results are listed in Table 2.1 through t_{50}. To the number of decimal places printed, the energy is constant and equal to $1/6$, and the example is complete.

2.4 Remarks

In practice, the leap frog formulas offer no subtle difficulties beyond the usual ones related to numerical accuracy and stability. These will be discussed later when specific applications are considered. This is not the case however with the conservative methodology described in Sect. 2.3. The first related problem is that the solution of systems like (2.26) and (2.27) need not be unique. This is because the system is implicit. To see this, consider, for example, (2.66) and (2.67), and set $k = 0$. Then elimination of v_1 implies

$$x_1^2 + x_1\left(x_0 - \frac{6}{h^2}\right) + \left(x_0^2 + \frac{6}{h^2}x_0 + \frac{6}{h}\right) = 0$$

or

$$x_1^2 + \left(1 - \frac{6}{h^2}\right)x_1 + \left(1 + \frac{6}{h^2} + \frac{6}{h}\right) = 0 ,$$

the discriminant D of which is

$$D(h) = \left(1 - \frac{6}{h^2}\right)^2 - 4\left(1 + \frac{6}{h^2} + \frac{6}{h}\right) .$$

Hence $\lim_{h \to 0} D(h) = \infty$, so that for sufficiently small h, there exist two solutions. For example, for $h = 0.01$, the two solutions are $x_1 = 1.01005$, $x_2 = 59998$. Of course, only $x_1 = 1.01005$ is numerically and physically reasonable.

A second problem occurs when the zeros in the denominators of (2.27) are not removable. For example, in (2.66), $x_{k+1} - x_k$ divided into $\phi(x_{k+1}) - \phi(x_k)$, thus removing the singularity $x_{k+1} = x_k$. In cases where such a division is not possible,

Table 2.1

x	y	Energy
1.0000000000	1.0000000000	0.16667
1.0100505042	1.0101008418	0.16667
1.0202030372	1.0204057513	0.16667
1.0304596603	1.0309188827	0.16667
1.0408224772	1.0416444964	0.16667
1.0512936345	1.0525869623	0.16667
1.0618753232	1.0637507632	0.16667
1.0725697795	1.0751404983	0.16667
1.0833792864	1.0867608866	0.16667
1.0943061747	1.0986167710	0.16667
1.1053528241	1.1107131220	0.16667
1.1165216650	1.1230550415	0.16667
1.1278151790	1.1356477675	0.16667
1.1392359012	1.1484966777	0.16667
1.1507864211	1.1616072945	0.16667
1.1624693840	1.1749852893	0.16667
1.1742874929	1.1886364874	0.16667
1.1862435097	1.2025668731	0.16667
1.1983402570	1.2167825944	0.16667
1.2105806199	1.2312899687	0.16667
1.2229675471	1.2460954883	0.16667
1.2355040537	1.2612058258	0.16667
1.2481932220	1.2766278403	0.16667
1.2610382042	1.2923685837	0.16667
1.2740422236	1.3084353066	0.16667
1.2872085775	1.3248354652	0.16667
1.3005206384	1.3415767283	0.16667
1.3140418570	1.3586669843	0.16667
1.3277157637	1.3761143484	0.16667
1.3415659713	1.3939271707	0.16667
1.3555961773	1.4121140439	0.16667
1.3698101666	1.4306838116	0.16667
1.3842118136	1.4496455771	0.16667
1.3988050850	1.4690087122	0.16667
1.4135940429	1.4887828666	0.16667

Table 2.1 *cont.*

x	y	Energy
1.4285828471	1.5089779775	0.16667
1.4437757584	1.5296042798	0.16667
1.4591771414	1.5506723163	0.16667
1.4747914677	1.5721929490	0.16667
1.4906233193	1.5941773700	0.16667
1.5066773917	1.6166371137	0.16667
1.5229584976	1.6395840686	0.16667
1.5394715704	1.6630304906	0.16667
1.5562216680	1.6869890160	0.16667
1.5732139764	1.7114726753	0.16667
1.5904538143	1.7364949077	0.16667
1.6079466368	1.7620695763	0.16667
1.6256980396	1.7882109836	0.16667
1.6437137639	1.8149338879	0.16667
1.6619997010	1.8422535203	0.16667
1.6805618966	1.8701856026	0.16667

as, for example, for $\ddot{x} = \sin x$, the method is still conservative, but the Newtonian iteration should be modified to

$$x_{k+1}^{(0)} = x_k + C_1, \, v_{k+1}^{(0)} = v_k + C_2, \, C_1 \neq 0, \, C_2 \neq 0 \,.$$

It should be noted also that if one is concerned only with conservation of energy and not with other system invariants, formulas of higher order accuracy like those of Sect. 2.3 are available (La Budde and Greenspan (1976a, 1976b), Marciniak (1985)).

Note finally that if one assigns different time steps, say, h_i, to the particles P_i, then the energy conservation results of Sect. 2.3 are still valid, which is useful for stiff problems.

Part II

Qualitative Newtonian Modeling

3

Elastic Strings and Solitons

3.1 Introduction

In science and engineering, quantitative modeling is always to be preferred. However, quantitative particle simulation requires a degree of experience which is more readily developed if one has already acquired some appropriate physical intuition. Indeed, good intuition is a major asset in any mathematical endeavor. Our objective, then, at present is the development of such intuition through the development of several qualitative models. Quantitative modeling will be pursued in Part III.

3.2 Discrete Strings

In precomputer days, the study of vibrations of an elastic string from an initial position of tension was modeled mathematically by means of the wave equation (Haberman (1987)):

$$\frac{\partial^2 u}{\partial t^2} - c^2 \frac{\partial^2 u}{\partial x^2} = 0 \tag{3.1}$$

in which c is a nonzero constant. The linearity of (3.1) enables one to solve related fixed end problems by means of Fourier series. However, equation (3.1) is *not* the equation for "large" vibrations (Sokolnikoff and Redheffer (1966)), even though it has been applied to the study of large vibrations. At best, it applies to the study of small vibrations, like those inherent in steel strings. Let us then proceed to a particle model which can exhibit large vibrations, and for simplicity, attention will be directed to two dimensions only. Extension to three dimensions follows in a natural way.

A discrete string is one which is composed of a finite number of particles P_i, $i = 1, 2, \ldots, n$, with respective centers (x_i, y_i), as shown simplistically in Fig. 3.1. For $\Delta t > 0$, $t_k = k\Delta t$, $k = 0, 1, \ldots$, let P_j be a typical particle in motion, as shown in Fig. 3.2 for $2 \leq j \leq n - 1$. At time t_k, let P_{j-1}, P_j, P_{j+1} be located at $(x_{j-1,k}, y_{j-1,k})$, $(x_{j,k}, y_{j,k})$, and $(x_{j+1,k}, y_{j+1,k})$, respectively. Also, let $\vec{a}_{j,k}$, $\vec{v}_{j,k}$ be the acceleration and velocity, respectively, of P_j at t_k. At present, P_1 and P_n are fixed.

For the forces between particles, we proceed as follows. Let T_1 be the tensile force between P_{j-1} and P_j, while T_2 is the tensile force between P_j and P_{j+1} as shown in Fig. 3.2, for typical particle P_j. The total force $\vec{F}_{j,k} = (F_{j,x,k}, F_{j,y,k})$ at time t_k is taken to be

$$F_{j,x,k} = |T_2|\frac{x_{j+1,k} - x_{j,k}}{r_{j+1,j,k}} - |T_1|\frac{x_{j,k} - x_{j-1,k}}{r_{j,j-1,k}} - \alpha v_{j,x,k} \tag{3.2}$$

$$F_{j,y,k} = |T_2|\frac{y_{j+1,k} - y_{j,k}}{r_{j+1,j,k}} - |T_1|\frac{y_{j,k} - y_{j-1,k}}{r_{j,j-1,k}} - \alpha v_{j,y,k} - 980m_j, \tag{3.3}$$

in which m_j is the mass of P_j, α is a nonnegative damping factor, and

$$r_{j+1,j,k} = [(x_{j+1,k} - x_{j,k})^2 + (y_{j+1,k} - y_{j,k})^2]^{\frac{1}{2}} \tag{3.4}$$

$$r_{j,j-1,k} = [(x_{j,k} - x_{j-1,k})^2 + (y_{j,k} - y_{j-1,k})^2]^{\frac{1}{2}}. \tag{3.5}$$

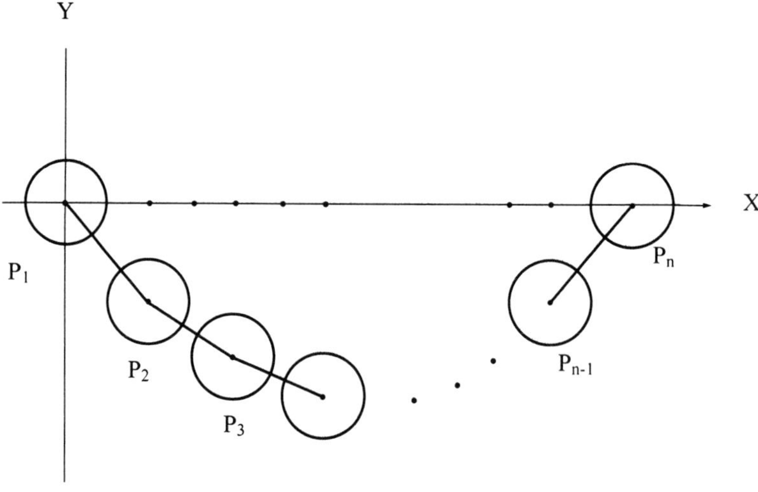

Figure 3.1.

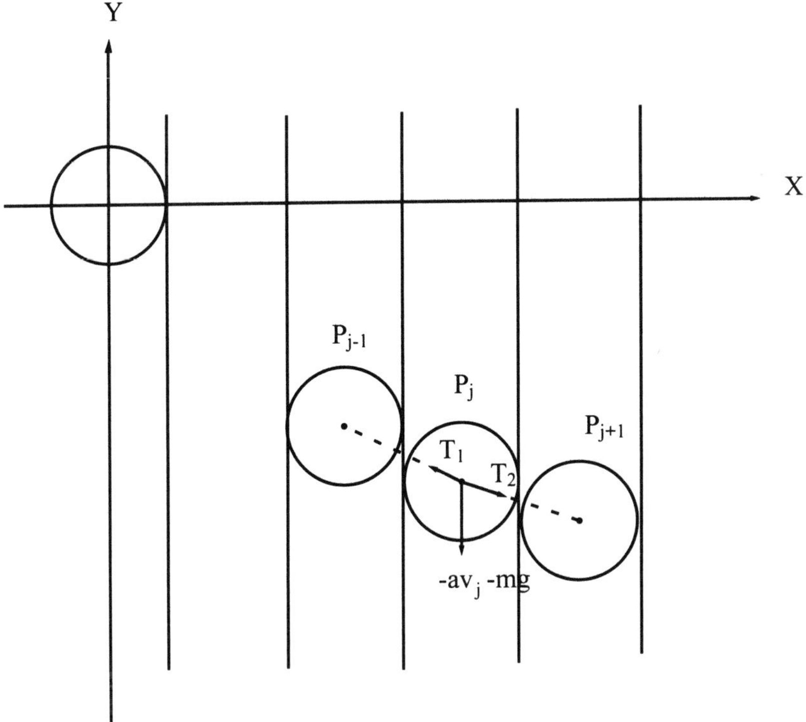

Figure 3.2.

The dynamical equations of motion are

$$m_j \vec{a}_{j,k} = \vec{F}_{j,k}, \qquad j = 2, \ldots, n-1 \tag{3.6}$$

$$k = 0, 1, 2, \ldots.$$

From given initial data, (3.2)–(3.6) can be solved readily by the leap frog formulas to yield the resulting vibrational motion.

3.3 Example

Let us proceed now to some simple string examples. In constructing these, the most difficult problem is decide upon the formulas for T_1 and T_2 in (3.2) and (3.3). The most common practice in choosing T_1 and T_2 is to assume that each varies directly with the distance between the respective particles involved. Thus, for example, if the distance between P_{j-1} and P_j in Fig. 3.2 is r_1, then one would assume

$$|T_1| = cr_1 \, , c > 0 \, , \tag{3.7}$$

where c is a constant which depends only on the physical nature of the string. Formula (3.7) is a particular example of what is called Hooke's law. Physical experimentation, however, reveals that Hooke's law is only of limited value. Indeed, (3.7) implies that if T_1 is arbitrarily large, so is r_1, and this simply is not correct. What happens, in fact, is that as T_1 continues to increase, r_1 increases less and less with T_1, until an elastic limit is reached and then the string actually breaks. Such qualitative behavior is nonlinear, while (3.7) is linear.

In our computer examples, then, we will try to simulate fully nonlinear behavior by suggesting and then using various possible formulas for T_1 and T_2 which may, or may not, include Hooke's law as a special case.

Consider a simple 21 particle string with $x_j = \frac{j-1}{10} \, , j = 1, 2, \ldots, 21$ and $\alpha = 0.15$. Assume that the two particles P_1 and P_{21} are fixed at $(0,0)$ and $(2,0)$, respectively. Further, assume that the moving particles $P_2, P_3, \ldots, P_{20}$ can move in the y direction only and that, at present, gravity is negligible. The resulting particle motions are called transverse and the string must converge to horizontal steady state from any initial configuration. In (3.2) and (3.3), let us set

$$\begin{cases} T_1 = T_0[1 + \left|\frac{y_{j,k} - y_{j-1,k}}{\Delta x}\right| + \frac{\epsilon}{2}\left|\frac{y_{j,k} - y_{j-1,k}}{\Delta x}\right|^2] \\ T_2 = T_0[1 + \left|\frac{y_{j+1,k} - y_{j,k}}{\Delta x}\right| + \frac{\epsilon}{2}\left|\frac{y_{j+1,k} - y_{j,k}}{\Delta x}\right|^2]. \end{cases} \tag{3.8}$$

Formulas (3.8) are simple nonlinear relationships which describe the tension between successive particles as a function of the slope of the segments joining the centers of these particles. The string is placed in an initial position as follows. P_2 is placed at $(0.1, 0.5)$ and P_3 at $(0.2, 1.0)$. The remaining particles are centered on $y = -\frac{5}{9}(x - 2)$, as shown at $t = 0$ in Fig. 3.3. For $T_0 = 12.5 \, , \epsilon = 0.01$, and $\Delta t = 0.00025$, the first 0.75 seconds of motion are shown typically in Fig. 3.3. The reflection of the resulting wave as it moves from left to right is seen from $t = 0.40$ to $t = 0.70$. In all figures after $t = 0.15$, one sees small trailing waves which follow the primary wave. After 6 seconds of motion, the particle motions are less than 0.005 from the horizontal steady state.

3.4 String Solitons

Nonlinear partial differential equations, like the Korteweg-deVries equation,

$$\frac{\partial \phi}{\partial t} + \sigma\phi\frac{\partial \phi}{\partial x} + \nu\frac{\partial^3 \phi}{\partial x^3} = 0$$

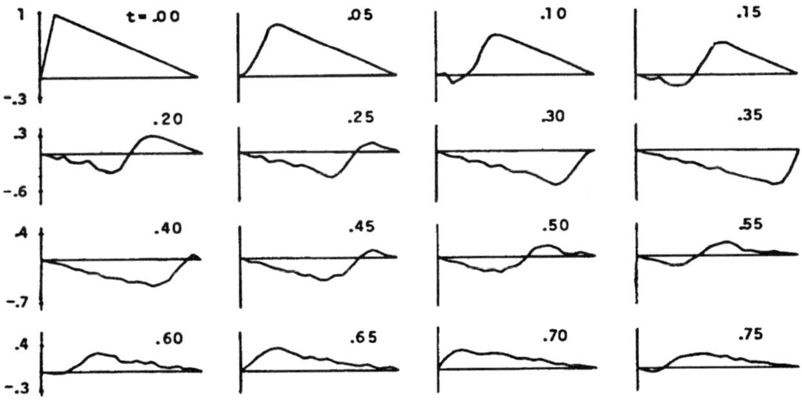

M o t i o n   o f   a   2 1   p a r t i c l e   s t r i n g

Figure 3.3.

in which σ and ν are constants, have solutions with properties shared by only a few other nonlinear equations. These particular solutions are isolated traveling waves which can pass through each other, unchanged in shape. These solutions are called solitons and now have a variety of physical applications. In the present section we will demonstrate soliton behavior by means of particle string modeling. In addition we will introduce a new tension formula which is also of interest.

In place of (3.8), let us consider

$$T_1 = T_0 \left\{ (1 - \epsilon)[\frac{[(x_{j,k} - x_{j-1,k})^2 + (y_{j,k} - y_{j-1,k})^2]^{\frac{1}{2}}}{x_{j,k} - x_{j-1,k}} \right. \tag{3.9}$$

$$\left. + \epsilon[\frac{[(x_{j,k} - x_{j-1,k})^2 + (y_{j,k} - y_{j-1,k})^2]^{\frac{1}{2}}}{x_{j,k} - x_{j-1,k}}]^2 \right\}$$

$$T_2 = T_0 \left\{ (1 - \epsilon)[\frac{[(x_{j+1,k} - x_{j,k})^2 + (y_{j+1,k} - y_{j,k})^2]^{\frac{1}{2}}}{x_{j+1,k} - x_{j,k}} \right. \tag{3.10}$$

$$\left. + \epsilon[\frac{[(x_{j+1,k} - x_{j,k})^2 + (y_{j+1,k} - y_{j,k})^2]^{\frac{1}{2}}}{x_{j+1,k} - x_{j,k}}]^2 \right\}$$

in which $0 \le \epsilon \le 1$. Consider 101 particles fixed initially as follows: $x_k = \frac{2(k-1)}{100}$, $k = 1, 2, \ldots, 101$, $y_k = 0, k = 1, 2, \ldots, 101$. Thus, all initial positions are known.

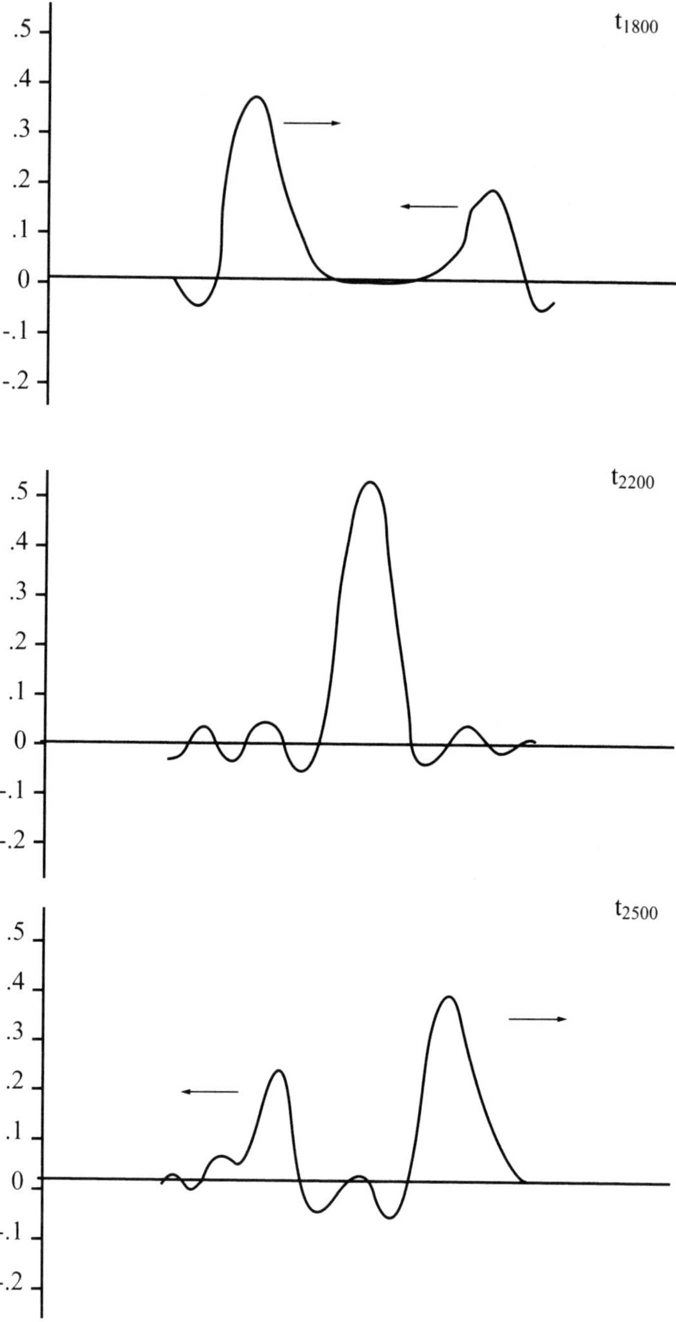

Figure 3.4.

Assume again that gravity can be neglected and that all particle motions are only transverse. Next, set $T_0 = 10$, $\alpha = 0.01$, $\Delta t = 0.0001$, $m = 0.01$, $\epsilon = 0.02$.

To generate two solitons, let us first fix P_1 at $(0,0)$ and P_{101} at $(2,0)$. Next set all velocities equal to zero except $v_{1,y,0} = 60$, $v_{2,y,0} = 50$, $v_{3,y,0} = 40$, $v_{4,y,0} = 30$, $v_{5,y,0} = 20$, $v_{6,y,0} = 10$, $v_{96,y,0} = 10$, $v_{97,y,0} = 20$, $v_{98,y,0} = 30$, $v_{99,y,0} = 40$. Application of the leap frog formulas then yields the soliton motion shown in Fig. 3.4.

3.5 Heavy Strings and Strings with One Fixed End

Particle string modeling extends directly to complex vibrational motions which are not readily accessible by continuum modeling. In this section we consider two such cases.

Consider first a heavy, elastic string, that is one whose equilibrium position is not horizontal because the effect of gravity is not negligible relative to the tensile forces (3.7), (3.8). As in Sect. 3.3, consider a 21-particle string for which P_1 and P_{21} are fixed while P_2, P_3, ..., P_{20} move in the y direction only. Let the initial particle configuration be that shown at $t = 0$ in Fig. 3.5. Then, for typical force parameters (Greenspan (1981a)), the downward swing is shown in Fig. 3.5 until $t = 0.35$, when the string begins its upward motion. The upward motion is shown in Fig. 3.6 from $t = 0.35$ to $t = 0.69$, at which time the string begins a downward motion. The equilibrium configuration is shown after six seconds in Fig. 3.7. At this time the motion is relatively negligible in that y changes in absolute value by at most 0.001.

Consider finally a heavy, 21-particle elastic string in which P_1 and P_{21} are fixed, initially, and have the configuration shown at $t = 0$ in Fig. 3.8. At $t = 0$, one then releases P_1 from its fixed position. For forces (3.9), (3.10) and with typical force parameters (Schubert and Greenspan (1972)), Fig. 3.8 shows the resulting motion from $t = 0$ to $t = 0.6$. Of course, it is necessary in the present problem to include particle motion in *both* the x and y directions, since, unlike the previous examples, motion in the x direction is not negligible relative to the motion in the y direction. In this case, the final equilibrium position is vertical, not horizontal.

3.6 Remark

For a particle model of a heavy rotating chain, see Snyman and Vermeulen (1979). For a discussion of standing waves, resonance, and stability, and for additional computations of particle strings, see Auret and Snyman (1977) and Gotusso and Veneziani (1994).

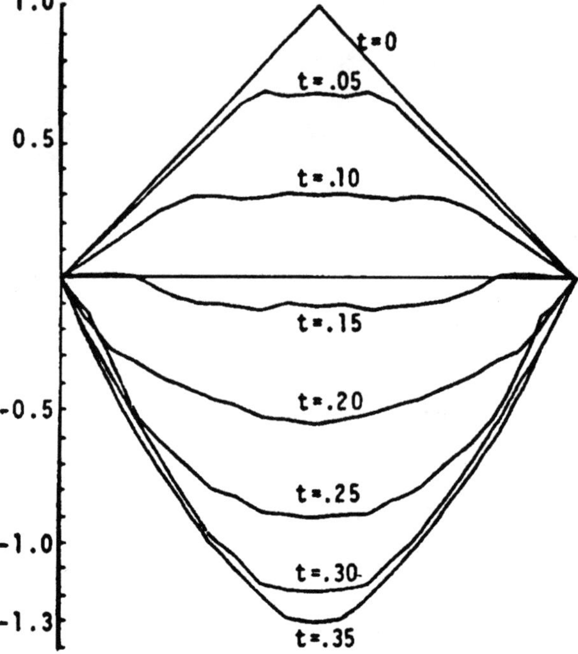

Figure 3.5.

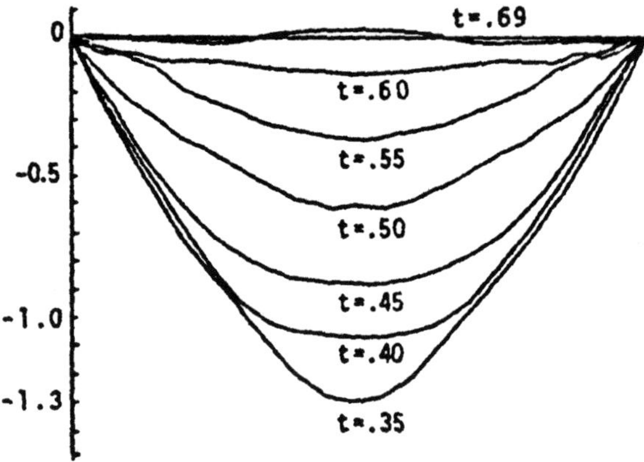

Figure 3.6.

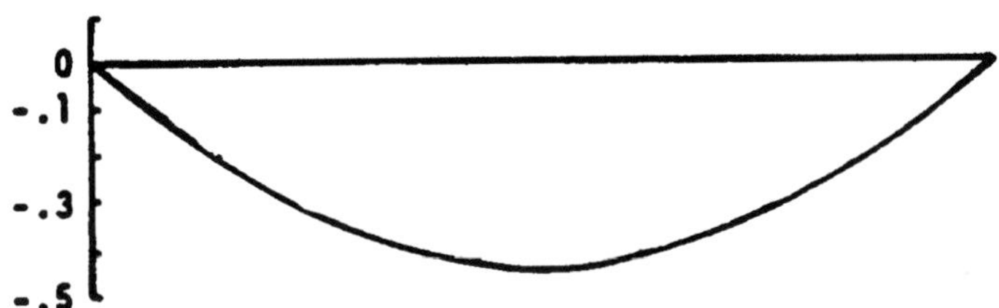

Figure 3.7.

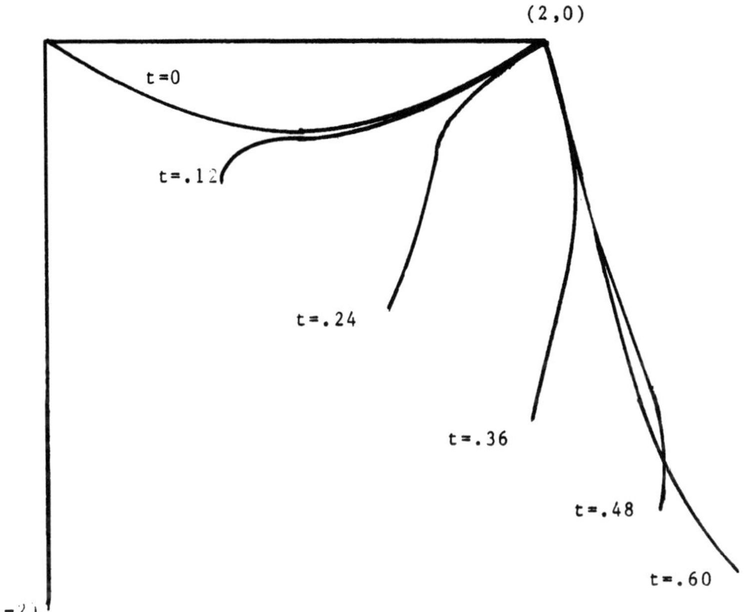

Figure 3.8.

4

Elastic Snap Through

4.1 Introduction

The response of an elastic material to stress is of broad scientific and engineering interest. Chapter 3 was concerned with a specific problem of this type. Another such problem is the spherical cap snap through problem (Keller and Reiss (1959)). In this problem one wishes to study the motion of a spherical, elastic cap, like half a tennis ball, when pressure is applied to its surface in order to make it invert, or turn inside out. Of fundamental interest in this problem is the cap's position of unstable equilibrium.

In this chapter we will simulate snap through, but, for simplicity, will restrict attention to two dimensions only. Instead of a cap, we will consider an arch. The resulting particle model will yield the position of unstable equilibrium in a direct, computational fashion.

4.2 An Arch

Consider, for example, 91 particles $P_1, P_2, P_3, \ldots, P_{91}$, each of unit mass, whose positions in the XY plane are shown as an arch in Fig. 4.1 and given precisely in Table 4.1. By construction, the particles are arranged on three concentric circular arcs of respective radii 18.2275, 19.1000, 19.9595, with the common center $(0, -13.506)$. The particles of the inner arc are numbered consecutively P_1–P_{30}, as shown, those of the middle arc P_{31}–P_{61}, and those of the outer arc P_{62}–P_{91}. The central angle subtended by each adjacent pair of particles on any arc is $3°$ and the distance from any particle P_i of the middle arc to the nearest particles in the two other arcs is unity. Finally, note that P_{31} and P_{61} lie on the X-axis, symmetrically about the Y axis. Next, the neighbors of every particle are defined as follows. The neighbors of P_1 are defined to be P_2, P_{31} and P_{32}. The neighbors of P_i, for $i = 2, 3, \ldots, 29$, are defined to be P_{i-1}, P_{i+1}, P_{i+30} and P_{i+31}. The neighbors of P_{30} are P_{29}, P_{60}, P_{61}. Those of P_{31} are P_1, P_{32} and P_{62}. For each of $i = 32, 33, \ldots, 60$, the neighbors of P_i are P_{i-31}, P_{i-30}, P_{i-1}, P_{i+1}, P_{i+30}, P_{i+31}. Those of P_{61} are P_{30}, P_{60} and P_{91}. The neighbors of P_{62} are defined to be P_{31}, P_{32} and P_{63}. Those of P_i, for $i = 63, 64, \ldots, 90$ are P_{i-31}, P_{i-30}, P_{i-1}, and P_{i+1}. Finally, the neighbors of P_{91} are taken to be P_{60}, P_{61} and P_{90}.

4.3 Elastic Snap Through

At time t_k, let the local force on P_i due to P_j be $\vec{F}_{ij,k}$. Assume that each P_i is acted upon *only by its neighbors* and that

$$\vec{F}_{ij,k} = \left[-\frac{1}{r_{ij,k}^2} + \frac{1}{r_{ij,k}^4} \right] \frac{\vec{r}_{ji,k}}{r_{ij,k}}.$$

Of course,

$$\vec{F}_{i,k} = m_i \vec{a}_{i,k},$$

in which $\vec{F}_{i,k}$ is the sum of the forces acting on P_i, and $r_{ij,k}$ is the distance between P_i and P_j.

The local forces are now defined completely. To simulate snap through inversion, however, one last assumption needs to be made. For simplicity, it is assumed that $\vec{F}_{31,k} \equiv \vec{F}_{61,k} \equiv \vec{0}, k = 0, 1, \ldots$, so that P_{31} and P_{61} will be fixed for all time. This assumption now allows us to neglect the effects of gravity.

With all initial velocities equal to $\vec{0}$ and with $\Delta t = 10^{-4}$, the leap frog formulas are now applied in the following way. After every 10,000 time steps, the Y

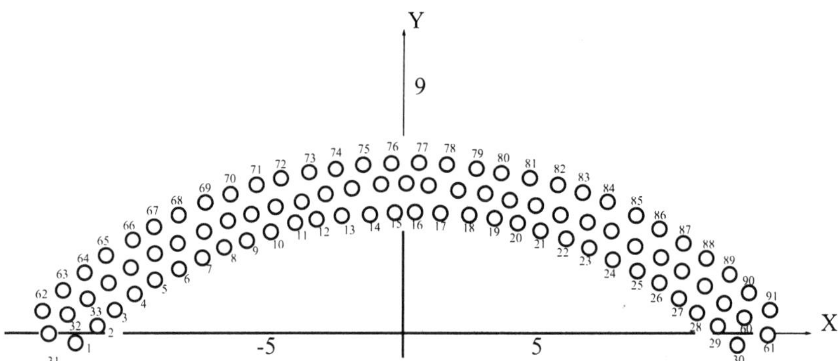

Figure 4.1 — Initial configuration.

Table 4.1   Coordinate Points of 91 Particles

i	x	y	i	x	y
1	−12.5470	−0.2842	47	0.9996	5.5678
2	−11.8378	0.3543	48	1.9965	5.4894
3	−11.0962	0.9548	49	2.9879	5.3588
4	−10.3242	1.5158	50	3.9711	5.1766
5	−9.5238	2.0355	51	4.9434	4.9432
6	−8.6974	2.5126	52	5.9022	4.6592
7	−7.8471	2.9459	53	6.8448	4.3254
8	−6.9754	3.3340	54	7.7687	3.9427
9	−6.0845	3.6760	55	8.6712	3.5122
10	−5.1769	3.9709	56	9.5500	3.0351
11	−4.2551	4.2179	57	10.4026	2.5126
12	−3.3217	4.4163	58	11.2267	1.9462
13	−2.3792	4.5656	59	12.0200	1.3375
14	−1.4301	4.6653	60	12.7804	0.6881
15	−0.4771	4.7153	61	13.5060	0.0000
16	0.4771	4.7153	62	−13.7390	0.9719
17	1.4301	4.6653	63	−12.9624	1.6710
18	2.3792	4.5656	64	−12.1503	2.3286
19	3.3217	4.4163	65	−11.3050	2.9429
20	4.2551	4.2179	66	−10.4286	3.5119
21	5.1769	3.9709	67	−9.5237	4.0344
22	6.0845	3.6760	68	−8.5926	4.5232
23	6.9754	3.3340	69	−7.6381	4.9338
24	7.8471	2.9459	70	−6.6625	5.3083
25	8.6974	2.5126	71	−5.6687	5.6312
26	9.5238	2.0355	72	−4.6593	5.9017
27	10.3242	1.5158	73	−3.6373	6.1189
28	11.0962	0.9548	74	−2.6052	6.2824
29	11.8378	0.3543	75	−1.5660	6.3916
30	12.5470	−0.2842	76	−0.5224	6.4463
31	−13.5060	0.0000	77	0.5224	6.4463
32	−12.7084	0.6881	78	1.5660	6.3916
33	−12.0200	1.3375	79	2.6052	6.2824
34	−11.2267	1.9462	80	3.6373	6.1189
35	−10.4026	2.5126	81	4.6593	5.9017
36	−9.5500	3.0351	82	5.6687	5.6312
37	−8.6712	3.5122	83	6.6625	5.3083
38	−7.7687	3.9427	84	7.6381	4.9338
39	−6.8448	4.3254	85	8.5926	4.5232
40	−5.9022	4.6592	86	9.5237	4.0344

Table 4.1 *cont.*

41	−4.9434	4.9432	87	10.4286	3.5119
42	−3.9711	5.1766	88	11.3050	2.9429
43	−2.9879	5.3588	89	12.1503	2.3286
44	−1.9965	5.4894	90	12.9624	1.6710
45	−0.9996	5.5678	91	13.7390	0.9719
46	0.0000	5.5940			

coordinates of the two particles P_{76} and P_{77} are both decreased by 0.1 to simulate a symmetrical downward push, or pressure, on the arch. In addition, at any time step, if the velocities in the Y direction of P_{76} and P_{77} are positive, these are reset to zero.

One further consideration must be introduced. Applied pressure adds to the total energy of the system, which unless dissipated, will result in large vibrational patterns. For this reason a damping factor δ is introduced, and after every 10,000 time steps, all velocities are multiplied by δ. For the present, δ is chosen as 0.98.

The computations were then carried out for 800,000 time steps, i.e., through $t = 80$. After $t = 80$ all pressure was released, the time step increased to $\Delta t = 0.002$ and the damping reset, correspondingly, to every 500 time steps. The entire motion and the resulting snap through inversion is shown in Figs. 4.2–4.12, at the respective times $t = 0, 20, 40, 60, 80, 100, 140, 180, 220, 880$. Figures 4.9–4.12 indicate the very rapid vibrations which result as the arch converges finally to a stable mode. It is possible that the curvature change, apparent in Figs. 4.7–4.9, causes these rapid vibrations, which, in turn, yields the sound associated with snap through.

For a second example, completely analogous, but nonsymmetrical, computations were made with only the following changes. The pressure was applied to P_{81} and P_{82}, rather than to P_{76} and P_{77}. In addition, not only were the Y coordinates of P_{81} and P_{82} decreased by 0.1 through $t = 80$, but the X coordinates were decreased by 0.01. All other aspects of the computations were identical. The resulting nonsymmetrical snap through is shown through $t = 220$ in Figs. 4.13–4.20. Thereafter, the configurations converge to the same steady state as indicated in Fig. 4.12 but in a rocking fashion.

4.4 Unstable Mode Approximation

Consider, again, the first example of the previous section. The maximum potential energy, −135.899, occurred at $t = 74$, so that one can expect to be close to an unstable mode at this time. However, to choose the $t = 74$ configuration might not be correct because the energy calculation is in error due to the impulsive motion

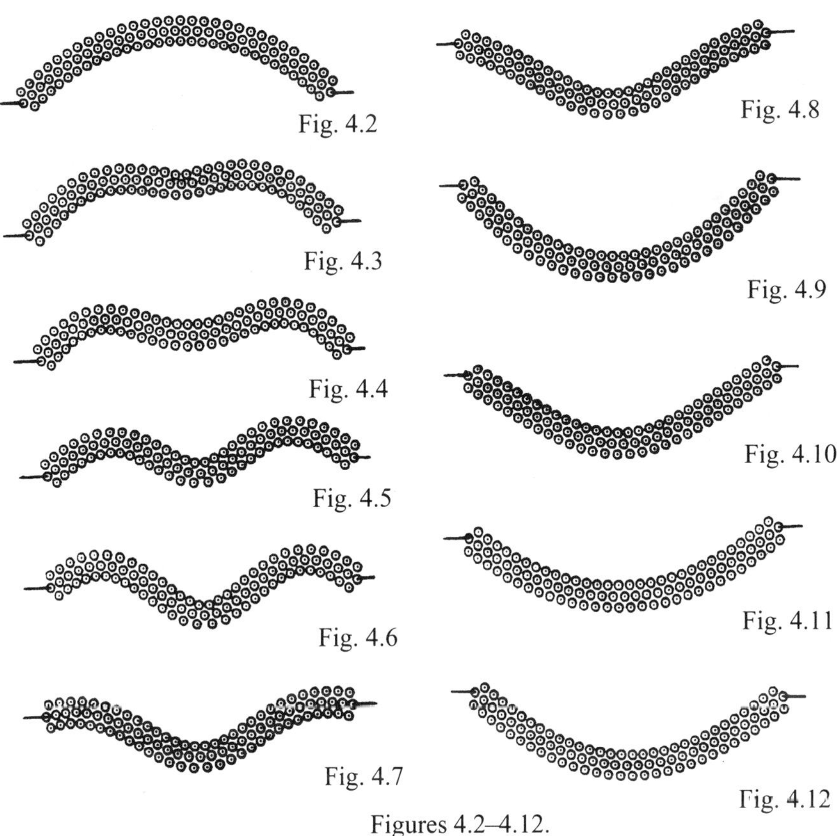

Fig. 4.2

Fig. 4.8

Fig. 4.3

Fig. 4.9

Fig. 4.4

Fig. 4.10

Fig. 4.5

Fig. 4.6

Fig. 4.11

Fig. 4.7

Fig. 4.12

Figures 4.2–4.12.

used to generate stress. Moreover, the use of damping was relatively arbitrary. In addition, the computer output revealed that, to one decimal place, the potential energy was -135.9 in the entire range $70 \leq t \leq 80$, so that any small error could easily shift the maximum potential energy to another time close to $t = 74$.

One way to proceed then, on physical grounds, is as follows. Consider the configuration S_1 which resulted at $t = 72.5$. Set each particle's velocity equal to zero and let S_1 vibrate without any applied pressure. The resulting net motion is upward. Then, using the configuration S_2 which resulted at $t = 77.5$, set each particle's velocity equal to zero and let S_2 vibrate without any applied pressure. The resulting net motion is downward. Physically, the unstable mode should be between S_1 and S_2 and it is approximated as follows. The point P_{46} moves from $(0, -1.81834)$ to $(0, 1.04666)$ in 20,000 time steps at $\Delta t = 0.005$ on S_1 and from $(0, -2.34466)$ to $(0, -2.89323)$ on S_2. Using linear interpolation it follows that P_{46} does not move at $t = 76.4$, which is taken to be the time of the unstable mode. The arch configuration at $t = 76.4$ is shown in Fig. 4.21.

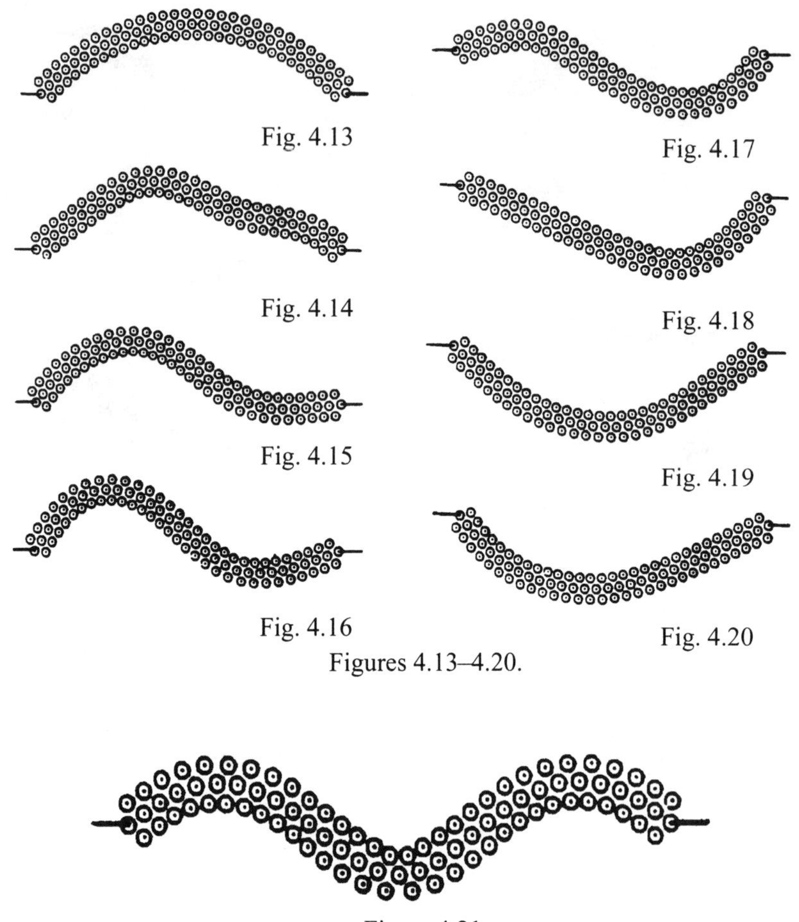

Fig. 4.13

Fig. 4.17

Fig. 4.14

Fig. 4.18

Fig. 4.15

Fig. 4.19

Fig. 4.16

Fig. 4.20

Figures 4.13–4.20.

Figure 4.21.

Finally, it is interesting to question the resulting motion, if any, of an unstable mode. Consider, then, the unstable mode shown in Fig. 4.21. If all pressure is released and all velocities set equal to zero, then the computational results show that, initially, the arch begins to exhibit small vibrations. These last a relatively long period. Thereafter, larger motions are generated and, eventually, the arch converges downward to a position of stable equilibrium. The implication is that any approximation will always have a small amount of error, that is, an exact unstable mode *does not exist*. However, this should not be disturbing because the same is true in Nature. In any physical arch, the molecules are always in random vibration and must, eventually, perturb an unstable mode into motion toward a stable mode. The computational results are therefore consistent with natural events, and this is so because molecular type forces have been used.

4.5 Remarks

With regard to the variety of other calculations performed, two results were of special interest. First, small variations of the damping parameter δ yielded dramatically different results. For example, for $\delta = 0.9$, the arch would barely move after pressure was removed, while for $\delta = 0.99$ the snap through proved to be so elastic that the arch dropped to a low point below -9. Second, changing the parameters $p = 2$, $q = 4$ to $p = 4$, $q = 6$ and decreasing appropriate Y coordinates 0.05 rather than 0.1 every 10,000 times steps resulted in configurations whose curvature near the fixed points P_{31} and P_{61} was relatively sharper. Other parameter changes yielded other expected qualitative results.

5

Minimal Surfaces

5.1 Introduction

Soap films and minimal surfaces have long been of interest both mathematically and physically (Almgren and Taylor (1976), Courant (1950), Radó (1951)). The Belgian physicist J.A.F. Plateau (1801–1883) determined experimentally a number of geometric properties of soap films by dipping a closed, thin wire into a soap solution and studying the resulting soap film, or minimal surface, which spanned the wire. Plateau concluded that every closed wire contour always bounds a minimal surface. Nevertheless, the nonuniqueness of the problem is now well known and extensive research has centered on proving the existence of at least one mathematical solution (Courant (1950)).

If one assumes sufficient differentiability of a particular class of soap films and assumes, in addition, that the film can be represented as a single valued function $u = u(x, y)$, then the Plateau problem can be formulated as a steady state boundary value problem for the nonlinear partial differential equation:

$$(1 + u_y^2)u_{xx} - 2u_x u_y u_{xy} + (1 + u_x^2)u_{yy} = 0, \tag{5.1}$$

or one in which the functional:

$$\iint\limits_{R \cup S} (1 + u_x^2 + u_y^2)^{\frac{1}{2}} dA \tag{5.2}$$

is to be minimized (Courant (1950), Radó (1951)). Functional (5.2) is the surface area of the film and equation (5.1) is the Euler equation of (5.2).

Analytically, only the most trivial problems can be solved using either equation (5.1) or functional (5.2). However, several computer methods have been developed for approximating solutions (Concus (1967), Greenspan (1967), Hinata, Shimasaki and Kiyono (1974)). These methods are of limited applicability because, in particular, they are restricted to single valued solutions. In this chapter, then, we

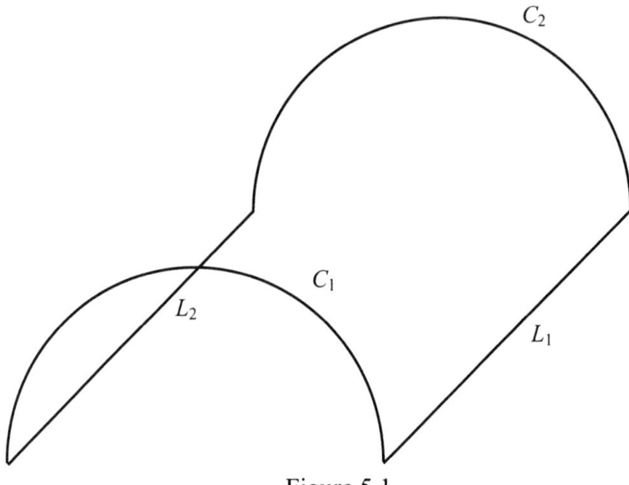

Figure 5.1.

will consider particle modeling of minimal surfaces, for which single valuedness need not be required.

5.2 Computer Examples

Consider first a class of nontrivial Plateau problems in which the boundary curves are of the three-dimensional type shown in Fig. 5.1, where C_1 and C_2 are parallel, semicircular segments of equal radii and L_1 and L_2 are parallel straight line segments of equal length.

Let us begin by considering first a rectangular array of only 43 particles P_1–P_{43}, arranged in a triangular mosaic as shown in Fig. 5.2. The distance between any two adjacent particles in the horizontal direction is taken to be 1.00645, while the distance between any two adjacent rows of particles is taken to be 0.866. The particles P_{10}–P_{17}, P_{19}–P_{25} and P_{27}–P_{34} are called the interior particles while the others are called the boundary particles. By construction, the interior particles P_{10}, P_{17}, P_{27}, and P_{34} have five neighbors each, while the remaining interior particles have six neighbors each.

The arrangement shown in Fig. 5.2 is then folded into a right cylindrical surface, as shown in Fig. 5.3. The particles P_1–P_9 are on C_1; P_{35}–P_{43} are on C_2; P_1, P_{18} and P_{35} are on L_1; and P_9, P_{26} and P_{43} are on L_2. The radii of both C_1 and C_2 are 2.56292 and the Euclidean distance between any interior particle and each of its neighbors is unity. Spatial coordinates of all 43 particles are given in Table 5.1 as initial particle positions.

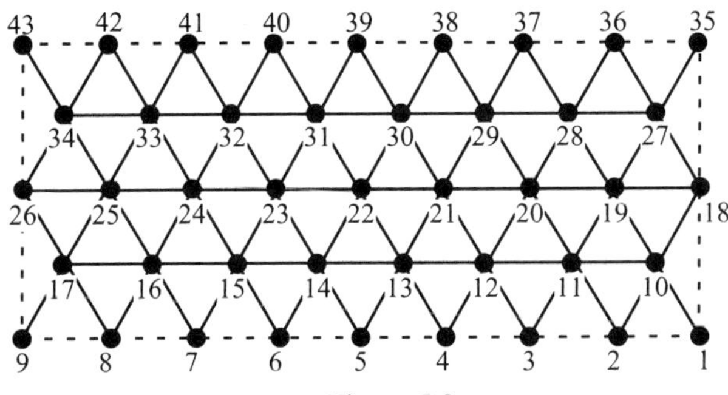

Figure 5.2.

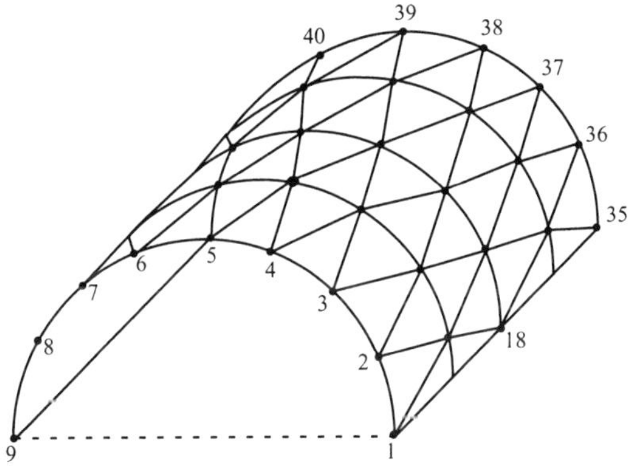

Figure 5.3.

The force parameter choices are $p = 2$, $q = 4$, $G = 1.0$, $H = 0.5$, so that $F = -\frac{1}{r_{ij}^2} + \frac{0.5}{r_{ij}^4}$. Gravity is neglected. The time step is $\Delta t = 0.001$. The boundary particles are fixed for all time and the interior particles are assigned zero initial velocities. Thereafter, the interior particles are allowed to vibrate to steady state by application of the leap frog formulas, with the convergence hastened by resetting all velocities equal to zero every 10,000 time steps. However, each particle is allowed to have a force interaction only with its neighbors.

The steady state minimal surface configuration is shown in Fig. 5.4 and the steady state positions are recorded in Table 5.1.

Consider next the more complex problem of approximating the soap film which spans two coaxial, disjoint rings. Such minimal surfaces have been the subject of many papers (see, e.g., Dickey (1966), Wagner (1977)). This particular problem is one for which the single valuedness assumption required for equation (5.1) is not valid.

Table 5.1

	Initial			Steady state		
i	x	y	z	x	y	z
1	0	2.563	0	0	2.563	0
2	0	2.368	0.981	0	2.368	0.981
3	0	1.812	1.812	0	1.812	1.812
4	0	0.981	2.368	0	0.981	2.368
5	0	0	2.563	0	0	2.563
6	0	−0.981	2.368	0	−0.981	2.368
7	0	−1.812	1.812	0	−1.812	1.812
8	0	−2.368	0.981	0	−2.368	0.981
9	0	−2.563	0	0	−2.563	0
10	0.866	2.514	0.500	0.885	2.283	0.651
11	0.866	2.131	1.424	0.928	1.851	1.305
12	0.866	1.424	2.131	0.949	1.215	1.831
13	0.866	0.500	2.514	0.957	0.423	2.125
14	0.866	−0.500	2.514	0.957	−0.423	2.125
15	0.866	−1.424	2.131	0.949	−1.215	1.831
16	0.866	−2.131	1.424	0.928	−1.851	1.305
17	0.866	−2.514	0.500	0.885	−2.283	0.651
18	1.732	2.563	0	1.732	2.563	0
19	1.732	2.368	0.981	1.732	2.059	0.907
20	1.732	1.812	1.812	1.732	1.505	1.518
21	1.732	0.981	2.368	1.732	0.800	1.921
22	1.732	0	2.563	1.732	0	2.063
23	1.732	−0.981	2.368	1.732	−0.800	1.921
24	1.732	−1.812	1.812	1.732	−1.505	1.518
25	1.732	−2.368	0.981	1.732	−2.059	0.907
26	1.732	−2.563	0	1.732	−2.563	0
27	2.598	2.514	0.500	2.579	2.283	0.651
28	2.598	2.131	1.424	2.536	1.851	1.305
29	2.598	1.424	2.131	2.515	1.215	1.831
30	2.598	0.500	2.514	2.507	0.423	2.125
31	2.598	−0.500	2.514	2.507	−0.423	2.125
32	2.598	−1.424	2.131	2.515	−1.215	1.831
33	2.598	−2.131	1.424	2.536	−1.851	1.305
34	2.598	−2.514	0.500	2.579	−2.283	0.651
35	3.464	2.563	0	3.464	2.563	0
36	3.464	2.368	0.981	3.464	2.368	0.981
37	3.464	1.812	1.812	3.464	1.812	1.812
38	3.464	0.981	2.368	3.464	0.981	2.368
39	3.464	0	2.563	3.464	0	2.563
40	3.464	−0.981	2.368	3.464	−0.981	2.368
41	3.464	−1.812	1.812	3.464	−1.812	1.812
42	3.464	−2.368	0.981	3.464	−2.368	0.981
43	3.464	−2.563	0	3.464	−2.563	0

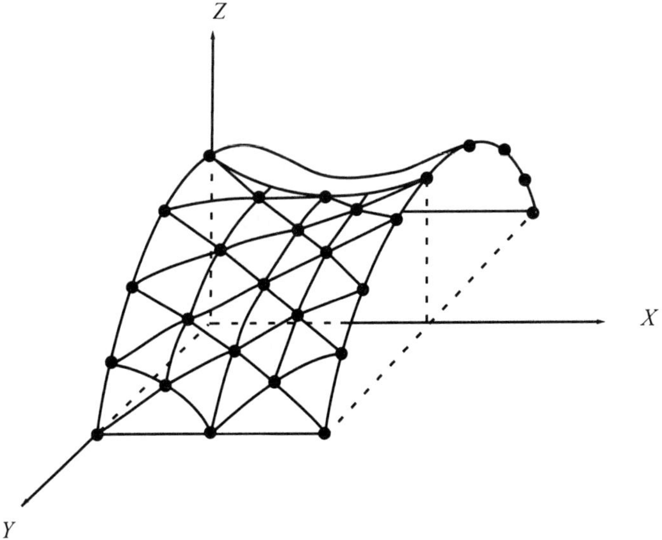

Figure 5.4.

We begin by constructing a cylindrical surface made up of particles arranged at the vertices of equilateral triangles, as shown in Fig. 5.5. The surface so constructed will consist of seven rings with 16 particles in each ring. The end rings represent the soap film which adheres to the two coaxial rings. Specifically, let $n = 112$ of which 32 particles represent the soap film particles attached to the end rings. The equilateral building blocks have sides of unit length. The 80 interior particles will be the only particles in vibration. All initial velocities are taken to be zero. The precise choice of initial positions is given in Table 5.2.

After some computer experimentation, the parameters chosen were $G = 0.87$, $H = 0.59$, $p = 2$, $q = 4$, $\Delta t = 0.001$. Thus, the leap frog formulas were implemented with

$$F_{ij} = -\frac{0.87}{r_{ij}^2} + \frac{0.59}{r_{ij}^4}.$$

Gravity was neglected and, this time, a local distance of interaction, $D = 1.31$, was introduced. Thus, for $|r_{ij}| \geq 1.31$, $F_{ij} = 0$. Symmetry allows the computation to be confined only to P_{17}, P_{33} and P_{49}. A type of damping less restrictive than that used for the result shown in Fig. 5.4 was implemented (Coppin and Greenspan (1988)). The resulting minimal surface is shown from two perspectives in Fig. 5.6 (a),(b). The final positions are given in Table 5.3.

Further examples in which the rings were gradually moved further apart led to a contraction of the neck, in agreement with experiment.

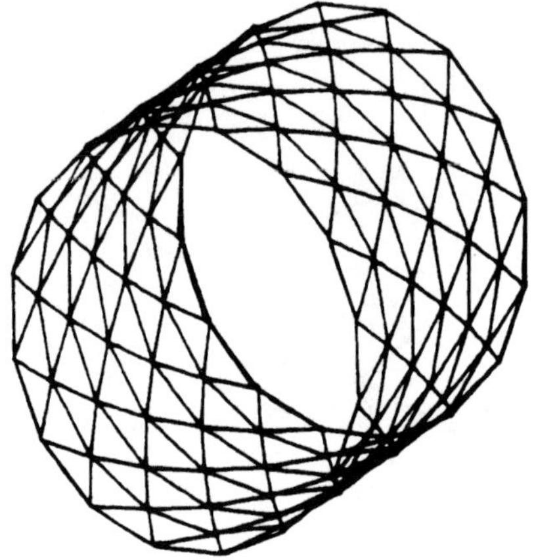

Figure 5.5.

Table 5.2   Initial Positions

Particle	x	y	z
1	−2.593872	−0.500000	2.513670
2	−2.593872	−1.423880	2.130986
3	−2.593872	−2.130986	1.423880
4	−2.593872	−2.513670	0.500000
5	−2.593872	−2.513670	−0.500000
6	−2.593872	−2.130986	−1.423880
7	−2.593872	−1.423880	−2.130986
8	−2.593872	−0.500000	−2.513670
9	−2.593872	0.500000	−2.513670
10	−2.593872	1.423880	−2.130986
11	−2.593872	2.130986	−1.423880
12	−2.593872	2.513670	−0.500000
13	−2.593872	2.513670	0.500000
14	−2.593872	2.130986	1.423880
15	−2.593872	1.423880	2.130986
16	−2.593872	0.500000	2.513670
17	−1.729248	0.000000	2.562915
18	−1.729248	−0.980785	2.367825
19	−1.729248	−1.812255	1.812255
20	−1.729248	−2.367825	0.980785
21	−1.729248	−2.562915	0.000000
22	−1.729248	−2.367825	−0.980785

Table 5.2   Initial Positions *cont.*

Particle	x	y	z
23	-1.729248	-1.812255	-1.812255
24	-1.729248	-0.980785	-2.367825
25	-1.729248	0.000000	-2.562915
26	-1.729248	0.980785	-2.367825
27	-1.729248	1.812255	-1.812255
28	-1.729248	2.367825	-0.980785
29	-1.729248	2.562915	0.000000
30	-1.729248	2.367825	0.980785
31	-1.729248	1.812255	1.812255
32	-1.729248	0.980785	2.367825
33	-0.864624	-0.500000	2.513670
34	-0.864624	-1.423880	2.130986
35	-0.864624	-2.130986	1.423880
36	-0.864624	-2.513670	0.500000
37	-0.864624	-2.513670	-0.500000
38	-0.864624	-2.130986	-1.423880
39	-0.864624	-1.423880	-2.130986
40	-0.864624	-0.500000	-2.513670
41	-0.864624	0.500000	-2.513670
42	-0.864624	1.423880	-2.130986
43	-0.864624	2.130986	-1.423880
44	-0.864624	2.513670	-0.500000
45	-0.864624	2.513670	0.500000
46	-0.864624	2.130986	1.423880
47	-0.864624	1.423880	2.130986
48	-0.864624	0.500000	2.513670
49	0.000000	0.000000	2.562915
50	0.000000	-0.980785	2.367825
51	0.000000	-1.812255	1.812255
52	0.000000	-2.367825	0.980785
53	0.000000	-2.562915	0.000000
54	0.000000	-2.367825	-0.980785
55	0.000000	-1.812255	-1.812255
56	0.000000	-0.980785	-2.367825
57	0.000000	0.000000	-2.562915
58	0.000000	0.980785	-2.367825
59	0.000000	1.812255	-1.812255
60	0.000000	2.367825	-0.980785

Table 5.2   Initial Positions *cont.*

Particle	x	y	z
61	0.000000	2.562915	0.000000
62	0.000000	2.367825	0.980785
63	0.000000	1.812255	1.812255
64	0.000000	0.980785	2.367825
65	0.864624	−0.500000	2.513670
66	0.864624	−1.423880	2.130986
67	0.864624	−2.130986	1.423880
68	0.864624	−2.513670	0.500000
69	0.864624	−2.513670	−0.500000
70	0.864624	−2.130986	−1.423880
71	0.864624	−1.423880	−2.130986
72	0.864624	−0.500000	−2.513670
73	0.864624	0.500000	−2.513670
74	0.864624	1.423880	−2.130986
75	0.864624	2.130986	−1.423880
76	0.864624	2.513670	−0.500000
77	0.864624	2.513670	0.500000
78	0.864624	2.130986	1.423880
79	0.864624	1.423880	2.130986
80	0.864624	0.500000	2.513670
81	1.729248	0.000000	2.562915
82	1.729248	−0.980785	2.367825
83	1.729248	−1.812255	1.812255
84	1.729248	−2.367825	0.980785
85	1.729248	−2.562914	0.000000
86	1.729248	−2.367825	−0.980785
87	1.729248	−1.812255	−1.812255
88	1.729248	−0.980785	−2.367825
89	1.729248	0.000000	−2.562925
90	1.729248	0.980785	−2.367825
91	1.729248	1.812255	−1.812255
92	1.729248	2.367825	−0.980785
93	1.729248	2.562915	0.000000
94	1.729248	2.367825	0.980785
95	1.729248	1.812255	1.812255
96	1.729248	0.980785	2.367825
97	2.593872	−0.500000	2.513670
98	2.593872	−1.423880	2.130986

Table 5.2   Initial Positions *cont.*

Particle	x	y	z
99	2.593872	−2.130986	1.423880
100	2.593872	−2.513670	0.500000
101	2.593872	−2.513670	0.500000
102	2.593872	−2.130986	−1.423880
103	2.593872	−1.423880	−2.130986
104	2.593872	−0.500000	−2.513670
105	2.593872	0.500000	−2.513670
106	2.593872	1.423880	−2.130986
107	2.593872	2.130986	−1.423880
108	2.593872	2.513670	−5.000000
109	2.593872	2.513670	5.000000
110	2.593872	2.130986	1.423880
111	2.593872	1.423880	2.130986
112	2.593872	0.500000	2.513670

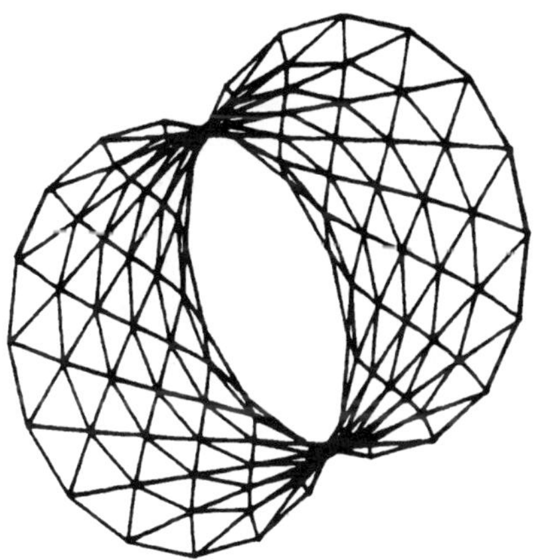

Figure 5.6(a).

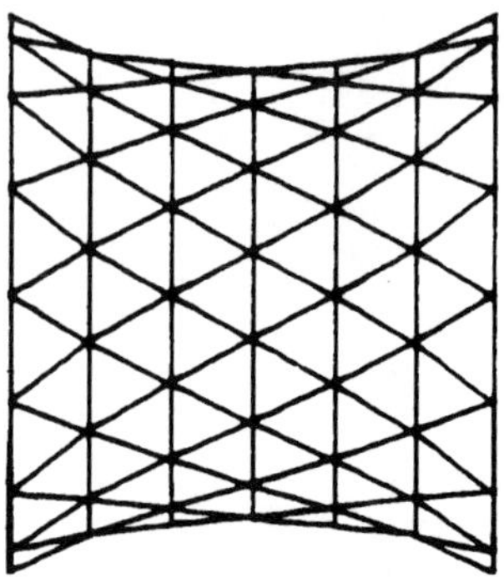

Figure 5.6(b).

Table 5.3   Final Positions

Particle	x	y	z
1	−2.593872	−0.500000	2.513670
2	−2.593872	−1.423880	2.130986
3	−2.593872	−2.130986	1.423880
4	−2.593872	−2.513670	0.500000
5	−2.593872	−2.513670	−0.500000
6	−2.593872	−2.130986	−1.423880
7	−2.593872	−1.423880	−2.130986
8	−2.593872	−0.500000	−2.513670
9	−2.593872	0.500000	−2.513670
10	−2.593872	1.423880	−2.130986
11	−2.593872	2.130986	−1.423880
12	−2.593872	2.513670	−0.500000
13	−2.593872	2.513670	0.500000
14	−2.593872	2.130986	1.423880
15	−2.593872	1.423880	2.130986
16	−2.593872	0.500000	2.513670
17	−1.755102	0.000000	2.254755
18	−1.755102	−0.862857	2.083122
19	−1.755102	−1.594353	1.594353
20	−1.755102	−2.083122	0.862857
21	−1.755102	−2.254755	0.000000
22	−1.755102	−2.083122	−0.862857
23	−1.755102	−1.594353	−1.594353
24	−1.755102	−0.862857	−2.083122

Table 5.3   Final Positions *cont.*

Particle	x	y	z
25	-1.755102	0.000000	-2.254755
26	-1.755102	0.862857	-2.083122
27	-1.755102	1.594353	-1.594353
28	-1.755102	2.083122	-0.862857
29	-1.755102	2.254755	0.000000
30	-1.755102	2.083122	0.862857
31	-1.755102	1.594353	1.594353
32	-1.755102	0.862857	2.083122
33	-0.882467	-0.414763	2.085152
34	-0.882467	-1.181144	1.767707
35	-0.882467	-1.767707	1.181144
36	-0.882467	-2.085152	0.414763
37	-0.882467	-2.085152	-0.414763
38	-0.882467	-1.767707	-1.181144
39	-0.882467	-1.181144	-1.767707
40	-0.882467	-0.414763	-2.085152
41	-0.882467	0.414763	-2.085152
42	-0.882467	1.181144	-1.767707
43	0.882467	1.767707	-1.181144
44	-0.882467	2.085152	-0.414763
45	-0.882467	2.085152	0.414763
46	-0.882467	1.767707	1.181144
47	-0.882467	1.181144	1.767707
48	-0.882467	0.414763	2.085152
49	0.000000	0.000000	2.093087
50	0.000000	-0.800990	1.933760
51	0.000000	-1.480036	1.480036
52	0.000000	-1.933760	0.800990
53	0.000000	-2.093087	0.000000
54	0.000000	-1.933760	-0.800990
55	0.000000	-1.480036	-1.480036
56	0.000000	-0.800990	-1.933760
57	0.000000	0.000000	-2.093087
58	0.000000	0.800990	-1.933760
59	0.000000	1.480036	-1.480036
60	0.000000	1.933760	-0.800990
61	0.000000	2.093087	0.000000
62	0.000000	1.933760	0.800990

Table 5.3   Final Positions *cont.*

Particle	x	y	z
63	0.000000	1.480036	1.480036
64	0.000000	0.800990	1.933760
65	0.882467	−0.414763	2.085152
66	0.882467	−1.181144	1.767707
67	0.882467	−1.767707	1.181144
68	0.882467	−2.085152	0.414763
69	0.882467	−2.085152	−0.414763
70	0.882467	−1.767707	−1.181144
71	0.882467	−1.181144	−1.767707
72	0.882467	−0.414763	−2.085152
73	0.882467	0.414763	−2.085152
74	0.882467	1.181144	−1.767707
75	0.882467	1.767707	−1.181144
76	0.882467	2.085152	−0.414763
77	0.882467	2.085152	0.414763
78	0.882467	1.767707	1.181144
79	0.882467	1.181144	1.767707
80	0.882467	0.414763	2.085152
81	1.755102	0.000000	2.254755
82	1.755102	−0.862857	2.083122
83	1.755102	−1.594353	1.594353
84	1.755102	−2.083122	0.862857
85	1.755102	−2.254755	0.000000
86	1.755102	−2.083122	−0.862857
87	1.755102	−1.594353	−1.594353
88	1.755102	−0.862857	−2.083122
89	1.755102	0.000000	−2.254755
90	1.755102	0.862857	−2.083122
91	1.755102	1.594353	−1.594353
92	1.755102	2.083122	−0.862857
93	1.755102	2.254755	0.000000
94	1.755102	2.083122	0.862857
95	1.755102	1.594353	1.594353
96	1.755102	0.862857	2.083122
97	2.593872	−0.500000	2.513670
98	2.593872	−1.423880	2.130986
99	2.593872	−2.130986	1.423880
100	2.593872	−2.513670	0.500000

Table 5.3   Final Positions *cont.*

Particle	x	y	z
101	2.593872	-2.513670	-0.500000
102	2.593872	-2.130986	-1.423880
103	2.593872	-1.423880	-2.130986
104	2.593872	-0.500000	-2.513670
105	2.593872	0.500000	-2.513670
106	2.593872	1.423880	-2.130986
107	2.593872	2.130986	-1.423880
108	2.593872	2.513670	-0.500000
109	2.593872	2.513670	0.500000
110	2.593872	2.130986	1.423880
111	2.593872	1.423880	2.130986
112	2.593872	0.500000	2.513670

6

Biological Self Reorganization

6.1 Introduction

Self reorganization is a common phenomenon in the sciences, especially in chemistry and biology. In this chapter we will turn attention to a particular type of self reorganization called biological cell sorting. Motivation is derived from biological experiments in which separated tissues self reorganized (Antonelli, Rogers and Willard (1973), Greenspan (1981(b)), Matela and Fletterick (1980), Rogers and Goel (1978), Steinberg (1963)).

A popular approach to the mathematical modeling of cell sorting is that of Steinberg (1963), which is founded on cellular motility and differential adhesion hypotheses, which, in turn, are dominated by the principle of minimization of free energy. Computer implementation of this approach has been explored using diverse dynamical possibilities including special motility rules, extended zone effects, and interface tension. In all related computer simulations, the prototype problem considered was as follows. In the plane, consider a collection of cells which consists of two different types, say, A and B. The initial positions of the cells are selected at random within a square. Confining all motions to be within the given square and using only *local* cell interaction rules, one must induce the A cells to organize into a central, relatively circular core, while the B cells form a layer around this core.

In this chapter we will modify Steinberg's theory in a natural and suitable fashion, so that self reorganization follows readily in the desired fashion from a particle model. The free energy minimization principle will be replaced by the mechanics of classical molecular interaction. Both double and triple layer self reorganization will be described and discussed.

6.2 Computer Examples

Since, in general, particles do not adhere when in a gaseous state and are rigid when in a solid state, sorting can only occur in a liquid, or near liquid state. Relative to this observation, extensive preliminary calculations motivate the following selection of parameters. Set

$$p = 3, q = 5, G_{ij} = H_{ij} = 5m_i m_j, \tag{6.1}$$

so that

$$F_{ij} = -\frac{5m_i m_j}{r_{ij}^3} + \frac{5m_i m_j}{r_{ij}^5} \tag{6.2}$$

Thus, F_{ij} is chosen so that it depends not only on r_{ij}, but also on m_i, the mass of P_i. Let the local distance of interaction parameter be given by

$$D = 2.2 \tag{6.3}$$

so that $r_{ij} \geq 2.2$ implies $\vec{F}_{ij} = 0$. Then, if P_i is to be a liquid particle, the range of the speed v_i of P_i can be deduced for various values of m_i. In particular (Greenspan (1980(a)))

$$m_i = 2000 \text{ implies } 100 \leq v_i \leq 170 \tag{6.4}$$

$$m_i = 4000 \text{ implies } 90 \leq v_i \leq 160 \tag{6.5}$$

$$m_i = 10000 \text{ implies } 50 \leq v_i \leq 80 \tag{6.6}$$

The parameter choices throughout this chapter will be guided by (6.1)–(6.6).

Consider now the following specific example. Consider the square region in the XY plane whose vertices are $(16, 16)$, $(-16, 16)$, $(-16, -16)$, $(16, -16)$. In this region, construct a triangular mosaic of 1072 grid points in the following fashion, which, incidentally, is sufficiently general to allow the construction of both larger and smaller sets of such grid points. Set

$$x_1 = -25.0, \, y_1 = 25.0, \, x_{52} = -24.5, \, y_{52} = 24.0$$

$$x_{i+1} = 1 + x_i, \, y_{i+1} = 25.0; \, i = 1, 2, \ldots, 50$$

$$x_{i+1} = 1 + x_i, \, y_{i+1} = 24.0; \, i = 52, 53, \ldots, 100$$

$$x_i = x_{i-101}, \, y_i = -1 + y_{i-101}; \, i = 102, 103, \ldots, 2576.$$

The resulting 2576 points P_i, with respective coordinates (x_i, y_i), are the vertices of a triangular mosaic which fills the square whose vertices are $(25, 25)$, $(-25, 25)$, $(-25, -25)$, $(25, -25)$. To determine the 1072 such points which lie within and on the square whose vertices are $(16, 16)$, $(-16, 16)$, $(-16, -16)$, $(16, -16)$, we merely exclude those of the 2576 points which satisfy any one of $x_i > 16$, $x_i < -16$, $y_i > 16$, $y_i < -16$.

Next, fix the set A to have 304 particles, each with $m_i = 10,000$, and the set B to have 768 particles, each with $m_i = 2000$. Each particle is set at a distinct grid

Figure 6.1 — Initial data.

point as shown in Fig. 6.1 where the particles of set A are represented by circles, while those of set B are represented by triangles. The particles of set A have been distributed widely throughout the square.

Next, a velocity is assigned to each particle, by a random process, in one of the four directions N,S,E,W. For each particle in B, the speed is 150. For each particle in A, the speed is either 50 or 80, determined at random. The velocity of each particle is shown in Fig. 6.1 as a vector emanating from each particle's center. All initial data are now assigned. For a complete listing, see the Appendix of Greenspan (1988f).

Now, fix $\Delta t = 0.0001$ and let the system parameters be given by (6.1)–(6.3). In order to keep the particles within the square while they are in motion, the following reflection rules are applied:

(a)   if $x_i > 16$, reset $x_i \to 32 - x_i$, $v_{x,i} \to -.99v_{x,i}$, $v_{y,i} \to .99v_{y,i}$

(b)   if $x_i < -16$, reset $x_i \to -32 - x_i$, $v_{x,i} \to -.99v_{x,i}$, $v_{y,i} \to .99v_{y,i}$

(c)   if $y_i > 16$, reset $y_i \to 32 - y_i$, $v_{x,i} \to .99v_{x,i}$, $v_{y,i} \to -.99v_{y,i}$

(d)   if $y_i < -16$, reset $y_i \to -32 - y_i$, $v_{x,i} \to .99v_{x,i}$, $v_{y,i} \to -.99v_{y,i}$

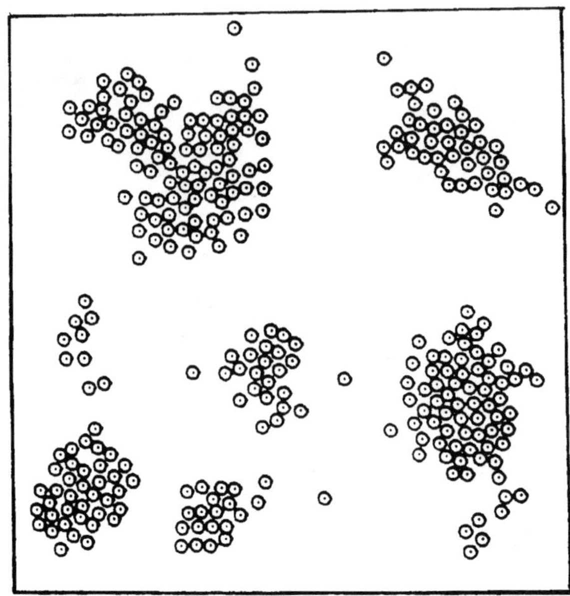

Figure 6.2 — T=0.5.

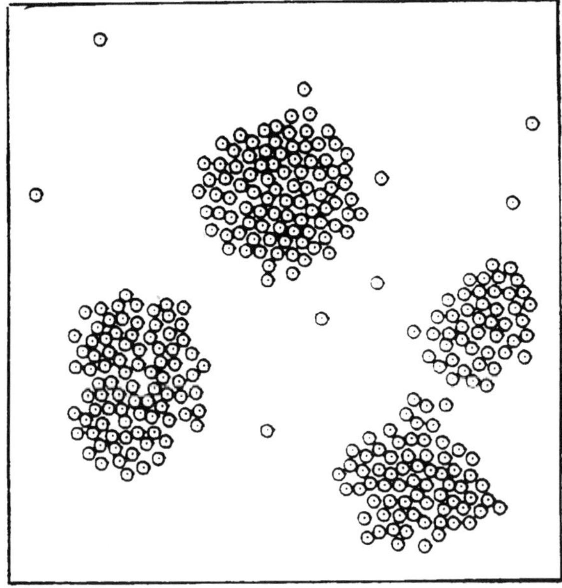

Figure 6.3 — T=1.5.

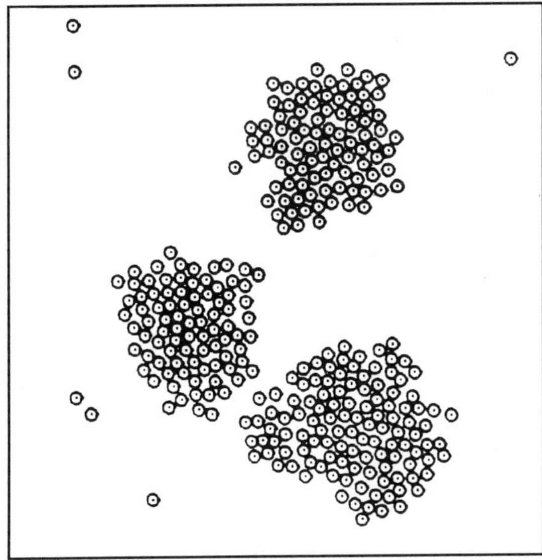

Figure 6.4 — T=2.5.

The small amount of velocity damping in rules (a) (d) insures stability of the leap frog formulas when using $\Delta t = 0.0001$, which will be discussed in greater detail later.

Figures 6.2–6.5 show the self reorganization process of the A particles at the respective times $T = 0.5, 1.5, 2.4, 4.0$. They reveal that these particles first form into small subsets, which then form into larger subsets, and finally form a central core. Figures 6.6–6.8, at the respective times $T = 9.0, 16.5, 24.0$ show the self reorganization of the set A into a relatively circular core. At this point, it was observed that the formation of an outer layer by the B particles was a relatively slow process. Hence, as an economy move, we reduced the damping factor in rules (a)–(d) to 0.9 after $T = 24.0$. Figure 6.9 then shows, at $T = 31.5$, the circular central core of A particles and the layer of B particles around the core.

As a second example, let us consider a set which consists of *three* different types of particles A, B, and C, and show how to induce a self reorganization in which the A particles form a central core, the B particles form a layer around the A particles, and the C particles form a layer around the B particles. The biological analog would be the self reorganization into normal tissue of separated endoderm, mesoderm, and ectoderm cells. All considerations are the same as in the first example with the following exceptions. The sets A, B, C have 38, 246 and 768 particles, respectively, with m_i values 10000, 4000 and 2000, respectively. The particles are positioned within the 32 x 32 square so that the A and the B sets have been widely separated. The A particles are each assigned speeds of 60 while the B and C particles are assigned speeds of 150. The initial data are displayed in Fig. 6.10, with the A particles represented by circles, the B particles

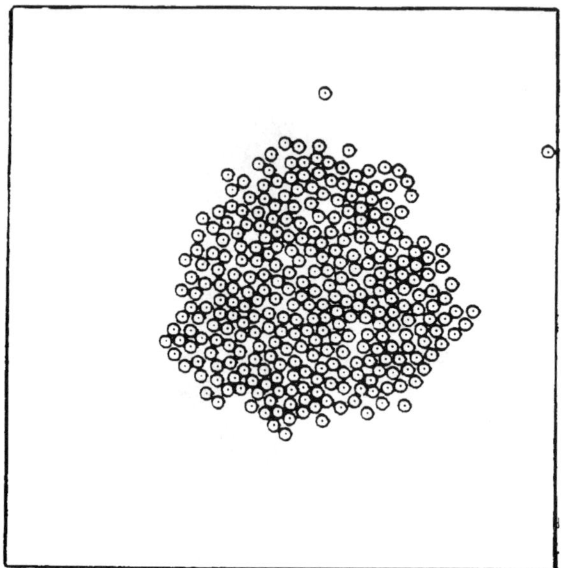

Figure 6.5 — T=4.0.

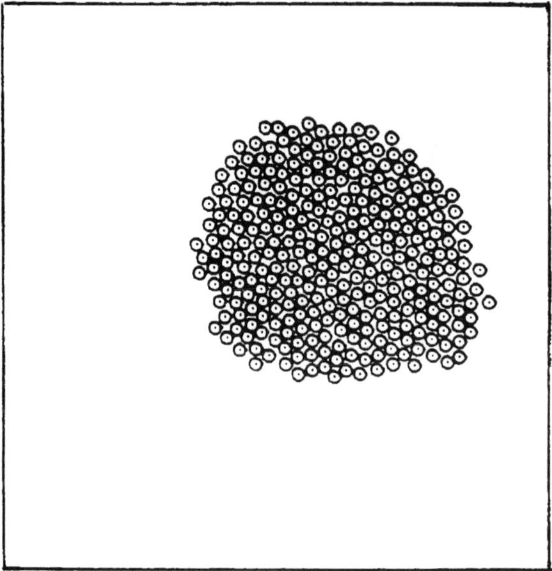

Figure 6.6 — T=9.0.

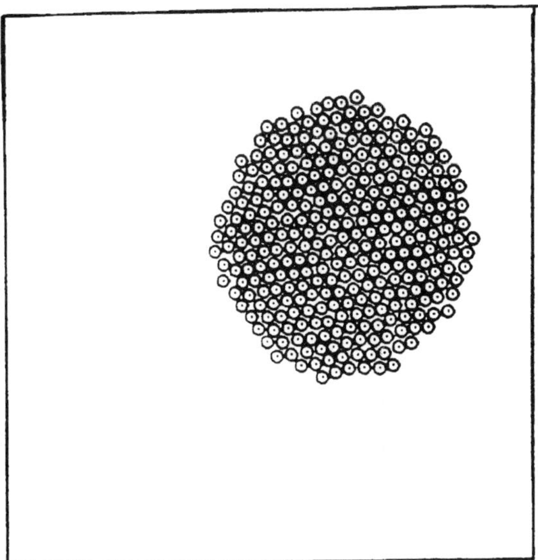

Figure 6.7 — T=16.5.

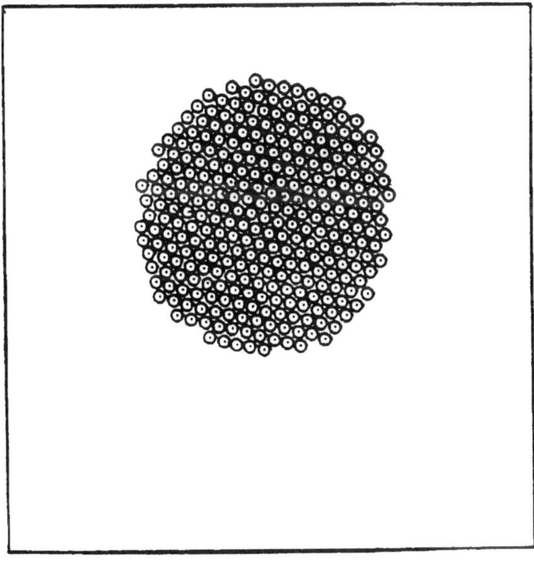

Figure 6.8 — T=24.0.

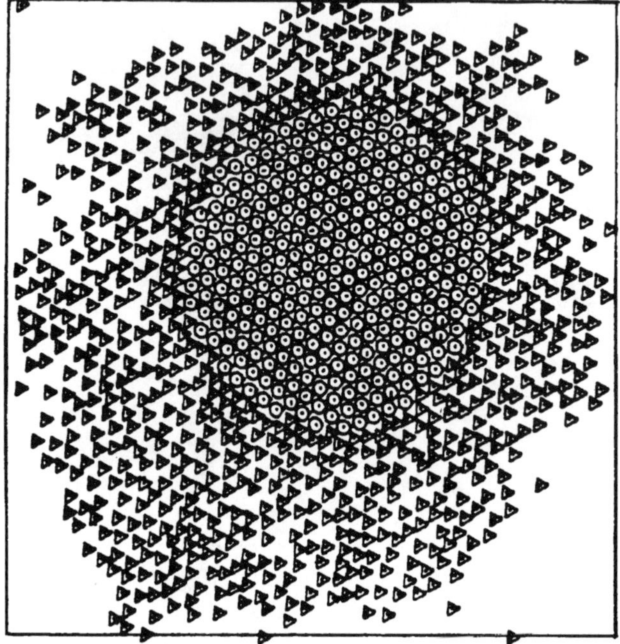

Figure 6.9 — T=31.5.

by quadrilaterals, and the C particles by triangles. A complete initial data listing is given in the Appendix of Greenspan (1988*f*).

The resulting self reorganization is shown in Figs. 6.11–6.20. Figures 6.11-6.16 show the self reorganization of the A cells at the respective times $T = 1.5$, 9.0, 16.5, 24.0, 31.5, 39.0. Figures 6.17–6.19 show the self reorganization of the B particles around the core at the respective times $T = 24.0$, 31.5, 39.0. Figure 6.20 shows the triple self reorganization of the A, B, C sets at time $T = 39.0$. The exceptionally slow self reorganization of sets B and C, after the A particles formed into the core was again accelerated by setting damping factor 0.99 to 0.9 in rules (a)–(d) after $T = 24.0$.

6.3 Remarks

Though small parameter variations in the two examples in Sect. 6.2 yielded entirely analogous results, large variations often did not. From the large number of additional examples run, we now discuss some of the types of problems which were thereby encountered.

Figure 6.10 — Initial data.

In the first example of Sect. 6.2, changing D to 3.0 and eliminating all damping resulted in successful execution until, approximately, t_{40000}. The result at this time was entirely analogous to that shown in Fig. 6.5. However, allowing D to be 3 increased the force after t_{40000} on each particle in the now highly concentrated A set to the point that instability resulted after t_{40000}. There were three possible remedies to correct this. First, one could decrease D. Second, one could introduce a small amount of damping in order to decrease particle velocities, and hence decrease the system's total kinetic energy. Third, one could decrease the time step. Since decreasing the time step was unfeasible economically, both the first and second remedies were implemented, which is, in fact, how the parameters of the first example were actually determined.

Next note that if one allows the m_i values to be "close," like 10,000 and 9500, then self reorganization is an exceptionally slow process.

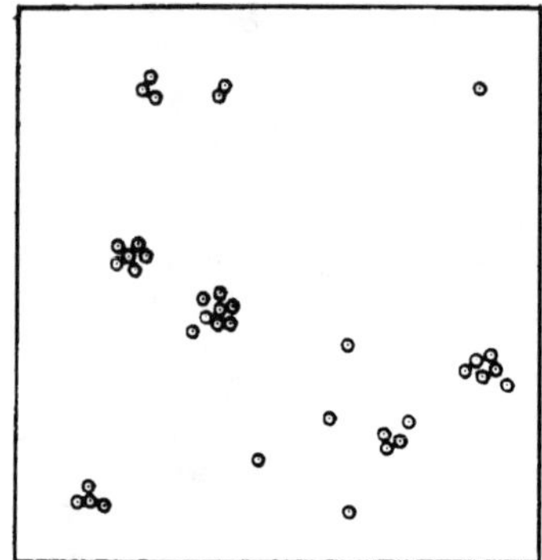

Figure 6.11 — T=1.5.

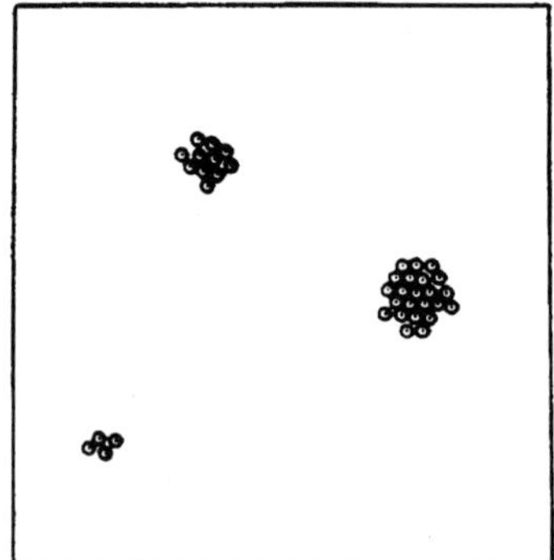

Figure 6.12 — T=9.0.

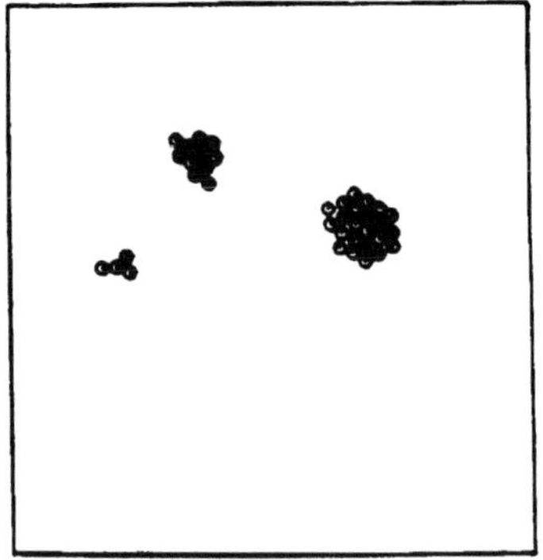

Figure 6.13 — T=16.5.

Figure 6.14 — T=24.0.

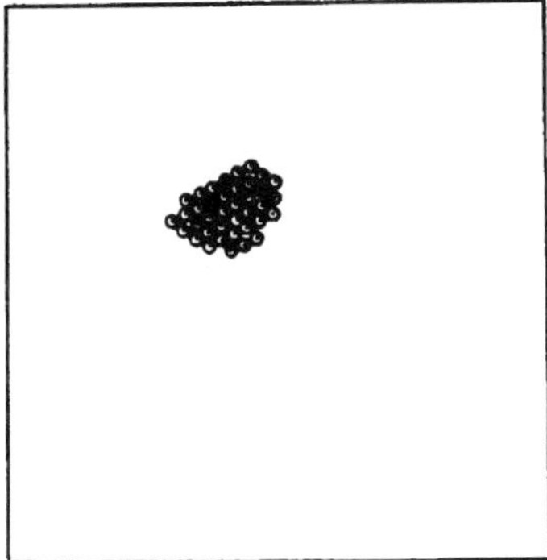

Figure 6.15 — T=31.5.

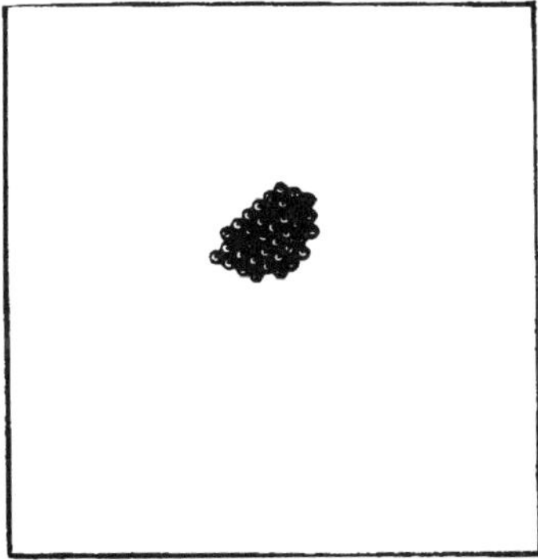

Figure 6.16 — T=39.0.

Figure 6.17 — T=24.0.

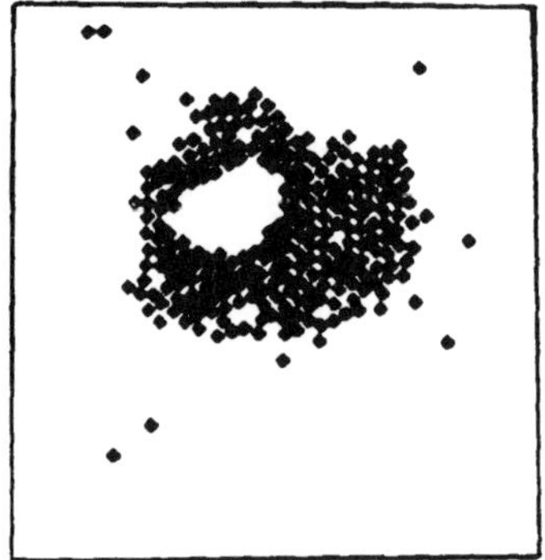

Figure 6.18 — T=31.5.

Note also that if, *from the start,* one chooses the damping factor in rules (a)–(d) to be too small, for example, 0.9, then *trapping* can result, that is, a particle from set B can always be found interior to set A in both examples of Sect. 6.2. The reason is that there follows an excessive loss of system kinetic energy, and this yields premature solidification. The biological implication is that sorting can occur only above a certain cell temperature T_1, which is characteristic of the cells under consideration.

Note also that for $D < 1.4$ and *without* damping, we were not be able to achieve sorting. This may have been due to the time constraints, or due to a system of kinetic energy level which was too high and yielded particle behavior like that of a gas, rather than that of a liquid. If the second possibility is correct, then there would also exist a temperature T_2, which is characteristic of the cells under consideration, *above* which sorting cannot result, and hence, sorting could only occur within the temperature range $T_1 < T < T_2$.

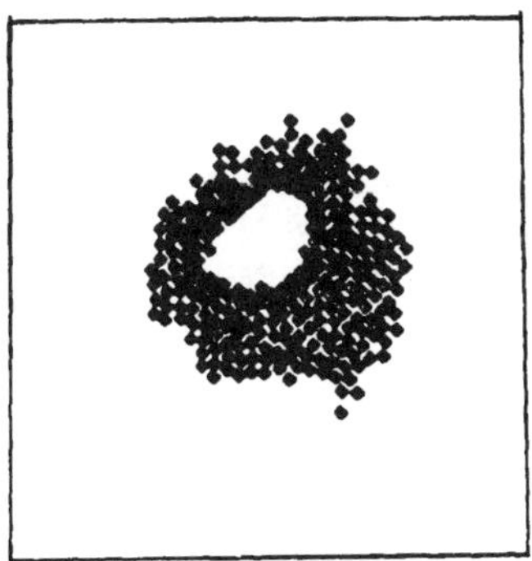

Figure 6.19 — T=39.0.

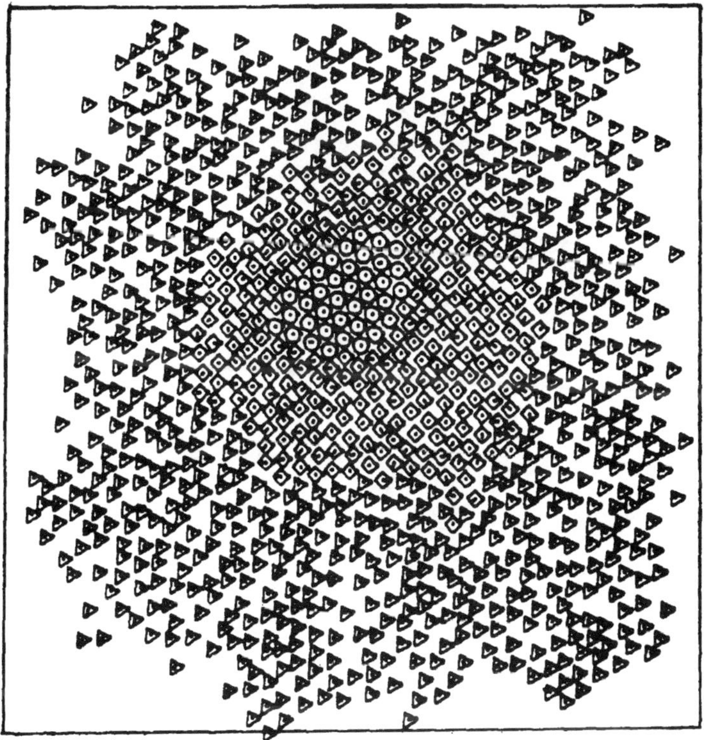

Figure 6.20 — T=39.0.

Finally, note that it is really somewhat ludicrous to interpret each m_i as a mass. Since we are using cgs units, $m_i = 10,000$ gr would be an unreasonable mass value. In the literature, then, each m_i is called an "adhesion constant," thus avoiding the problem.

All the remarks of this section should reenforce the opening observation that all the models discussed thus far are qualitative.

7

Cavity Flow

7.1 Introduction

Perhaps the simplest type of nontrivial, fluid dynamical problem which is used to test various types of models and numerical methods is the two-dimensional cavity flow problem. This problem is formulated as follows: Determine the motion of a fluid which fills a square basin, or cavity, when the upper side, or lid, is in uniform horizontal motion. In this chapter we will formulate and explore a particle model for the cavity problem.

7.2 Computer Example

Let us begin directly by describing in detail a specific example. For this purpose, consider a square $ABCD$ as shown in Fig. 7.1. The coordinates of the vertices are $A(-6.25, 6.25)$, $B(6.25, 6.25)$, $C(6.25, -6.25)$, $D(-6.25, -6.25)$. The inside of the square is called the basin and the top side AB is called the lid of the basin.

We now construct a triangular mosaic of 2576 points within and on $ABCD$, as shown in Fig. 7.1. The coordinates of the points are given by (x_i, y_i), where

$$x_1 = -6.25, \; y_1 = 6.25, \; x_{52} = -6.125, \; y_{52} = 6.0$$

$$x_{i+1} = x_i + 0.25, \quad y_{i+1} = 6.25; \quad i = 1, 2, \ldots, 50,$$

$$x_{i+1} = x_i + 0.25, \quad y_{i+1} = 6.0; \quad i = 52, 53, \ldots, 100,$$

$$x_i = x_{i-101}, \quad y_i = y_{i-101} - 0.5; \quad i = 102, 103, \ldots, 2576.$$

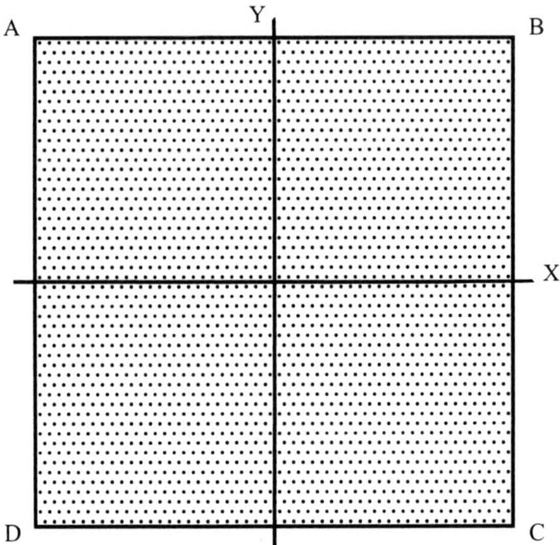

Figure 7.1 — The square cavity region.

This point set is symmetrical about both axes and the origin. The 51 rows contain, alternately, 51 and 50 points. In each row, the distance between two adjacent points is 0.25. The distance between two consecutive rows is also 0.25.

Each point will represent a particle and the mass of each particle is taken to be unity. The particle with coordinates (x_i, y_i) is denoted by P_i. Thus, the subscripts of the P_i increase from left to right on any row and the numbering begins on the top row and proceeds from any row to the next lower row.

The long-range force on each particle P_i is taken to be gravity, so that $g = 980.0$. Since all masses are unit masses, we take, as usual, the local force on P_i due to neighbor P_j to be

$$\vec{F}_{ij,k} = \left[-\frac{G}{(r_{ij,k})^p} + \frac{H}{(r_{ij,k})^q} \right] \frac{\vec{r}_{ji,k}}{r_{ij,k}}. \tag{7.1}$$

Now, since gravity will be a dominating force, local repulsion must be significant to keep all the particles from falling to the bottom of the basin. Hence, we now choose $G = 0$, $H = 100$, $p = 3$, $q = 5$. Assume, also, that local force interactions are restricted to pairs of particles whose distance of separation is less than $D = 0.35$.

Each particle is now assigned a small, randomly generated velocity vector whose speed is less than 0.002, so that all initial data are determined.

The resulting system of 2576 second order, dynamical equations was then solved numerically by the leap frog formulas with $\Delta t = 0.0001$. Whenever a

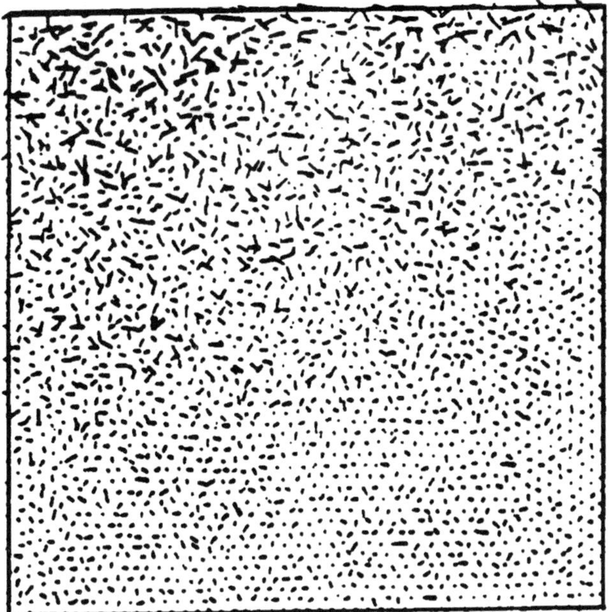

Figure 7.2 — Instantaneous velocity field at t_{6000}.

particle crossed a side of the square, it was reflected back symmetrically across that side with a velocity damping factor of 0.9. However, when a particle had moved across the top side AB of the square, that is, whenever the particle had collided with the lid of the square, the particle was reflected as indicated and then a constant V was added to its x-component of velocity. For the present, let $V = -10.0$.

In the usual notation $t_k = k\Delta t$, $k = 0, 1, 2, \ldots$, the system was solved by the leap frog formulas to t_{250000}. The results are described as follows.

Figure 7.2 shows the velocity field after 6000 time steps, that is, at t_{6000}. The figure is relatively meaningless because the field is dominated by Brownian type motions, which reflect the strong molecular type interactions. A smoothing or filtering process is therefore required to clarify the gross fluid motions, and this is implemented as follows.

DEFINITION 7.1 For J a positive integer, let particle P_i be at $(x_{i,k}, y_{i,k})$ at time t_k and at $(x_{i,k-J}, y_{i,k-J})$ at time t_{k-J}. Then P_i's average velocity $\vec{v}_{i,k,J}$ at time t_k is defined by

$$\vec{v}_{i,k,J} = \left(\frac{x_{i,k} - x_{i,k-J}}{J\Delta t}, \frac{y_{i,k} - y_{i,k-J}}{J\Delta t} \right).$$

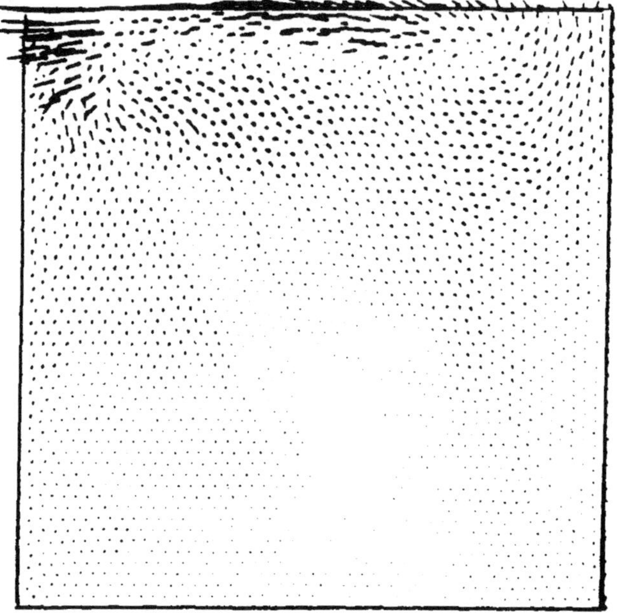

Figure 7.3 — Average velocity field at t_{1500}.

In this chapter, we choose $J = 1500$, which was decided upon after several comparison runs with other values of J. In the remainder of the discussion, all velocity fields are *average* velocity fields.

Figures 7.3–7.9 show the motion of the fluid by displaying the velocity fields at times in the range t_{1500}–t_{200000}. The development of a primary vortex is clear as is the motion of its core from the upper right to a more central location. These results are in agreement with experimental results (Pan and Acrivos (1967)). Most important, however, is the observation that the model reveals the mechanisms of vortex development. Indeed, Figs. 7.3–7.5 reveal that there is compression in the upper left corner and partial vacuum in the upper right corner. The compression yields large repulsive forces, which result in motion downward. The repulsive forces between particles just below the partial vacuum result in upward motion to fill the void. Thus, rotational motion begins. The continued driving force of the lid results in an increase in the size of the vortex. However, when the vortex has reached the size shown in Fig. 7.6 at t_{15000}, it changes relatively slowly thereafter. This is consistent with the usual assumption of a steady state (Anderson, Tannehill and Pletcher (1984), Pan and Acrivos (1967)), which results when the energy being added to the system is dissipated at the same rate.

The figures reveal also other interesting aspects of the flow. Figure 7.5 accentuates a downward flow near the left wall. Figures 7.4–7.9 show, with clarity,

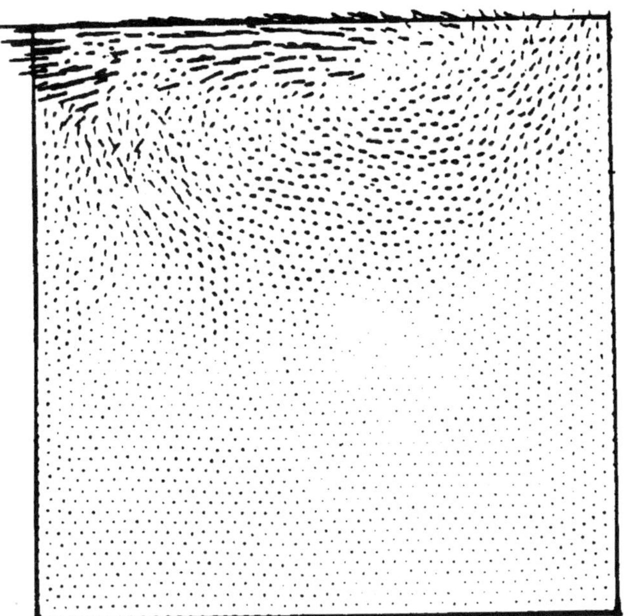

Figure 7.4 — Average velocity field at t_{3000}.

the development of "arms" at the bottom of the primary vortex which penetrate into the relatively quiescent fluid area below the vortex. Figures 7.7–7.9 accentuate the development of a relative dead zone near the right wall.

Since there is interest in the motion in the upper corners, we have shown in Figs. 7.10–7.13 at every 500 times steps from t_0 to t_{3000} the dispersive mixing of various upper left corner particles. Figure 7.10 shows the motion of P_1–P_4, Fig. 7.11 shows P_{52}–P_{55}, Fig. 7.12 shows P_{102}–P_{105}, and Fig. 7.13 shows P_{153}–P_{156}. Figures 7.10–7.13 reveal that P_1, P_{52}, P_{102}, and P_{153} exhibit small oscillations during this initial period, but remain relatively stationary. Figure 7.10 reveals an erratic motion for P_2, while Fig. 7.11 reveals large motions toward the interior for P_{53}–P_{55}. Figure 7.12 shows motion along the wall for P_{103}. Figure 7.13 reveals, initially, strong backward motions for P_{154} and P_{155}. It may be noted that if we had utilized a near neighbor algorithm like that of Boris (1986), we could not have described the individual motions of these particles.

Secondary vortices were not identified easily, primarily because a moving vortex is not readily recognizable and because small motions may require a different value of J for proper display. However, two examples of secondary vortex development are given in Figs. 7.14 and 7.15, one in the lower left corner at t_{150000} and one in the lower right corner at t_{200000}. For display purposes, the velocity vectors require magnification. If CTV represents the magnification factor, then CTV$= 25$ was used for Fig. 7.14 while CTV$= 15$ was used for Fig. 7.15.

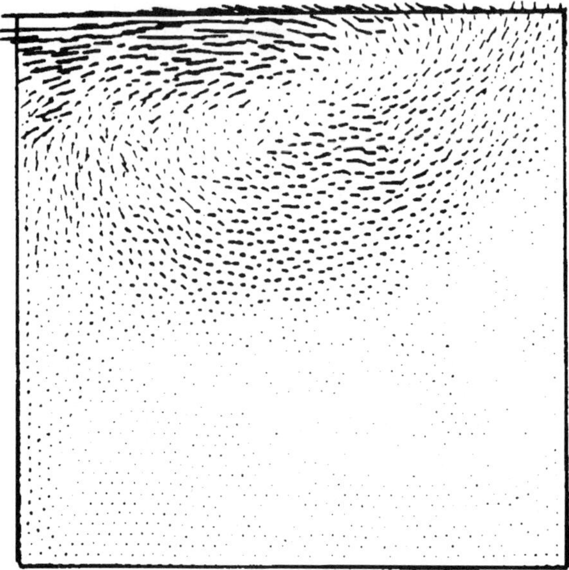

Figure 7.5 — Average velocity field at t_{6000}.

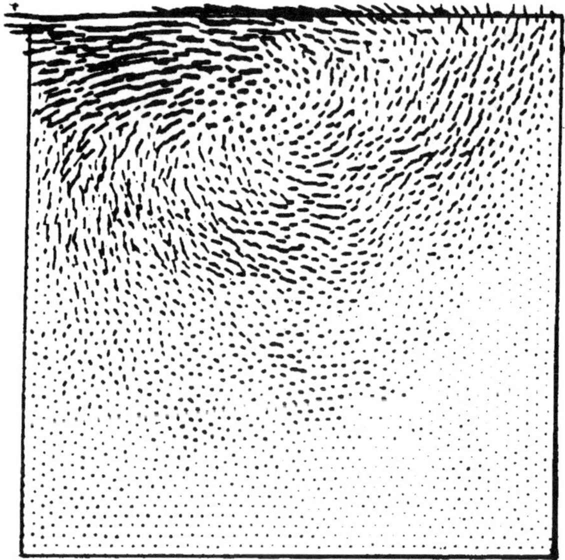

Figure 7.6 — Average velocity field at t_{15000}.

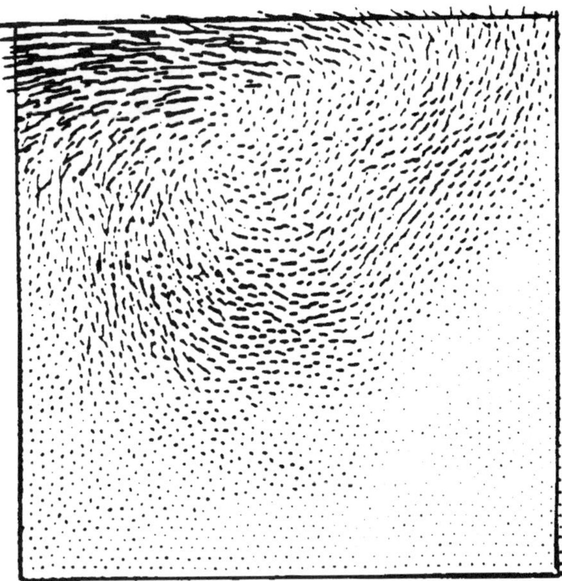

Figure 7.7 — Average velocity field at t_{35000}.

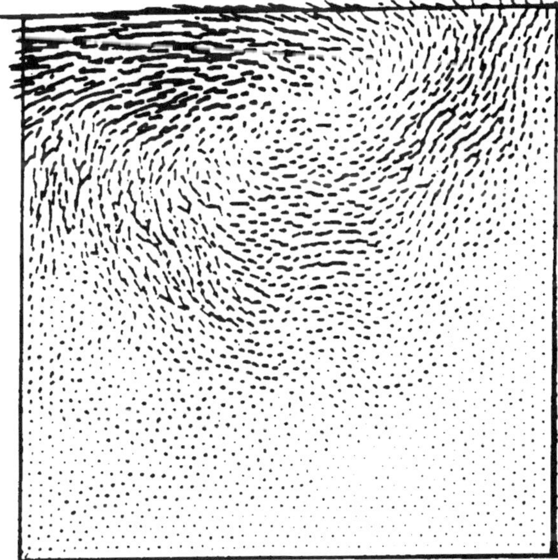

Figure 7.8 — Average velocity field at t_{100000}.

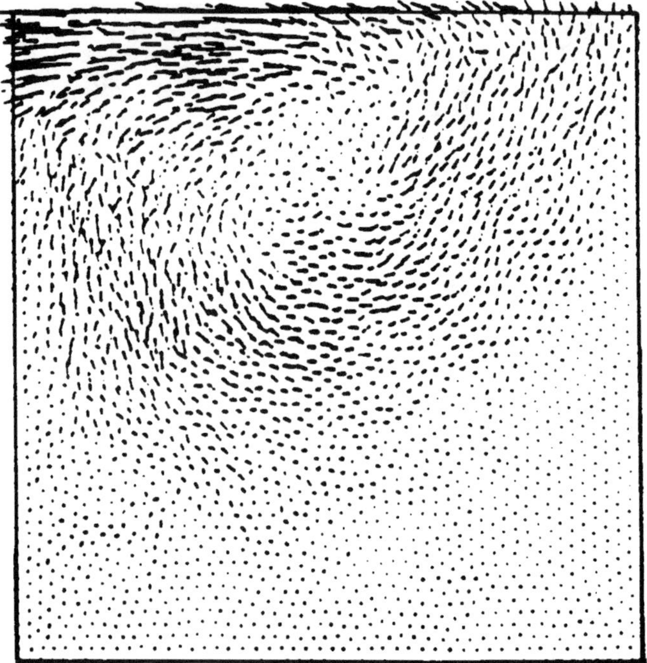

Figure 7.9 — Average velocity field at t_{200000}.

7.3 Additional Examples

The number and variety of examples which can be explored are, of course, unlimited. We therefore describe only three of these studies in which V was varied.

For $V = -2.5$, no vortex ever develops. Instead, there is an undulation through the fluid which resembles a compression wave. For the cases $V = -7$, $V = -13$, the results at t_{15000} are shown in Figs. 7.16 and 7.17, respectively. Relative to the results for $V = -10$ shown in Fig. 7.6, Fig. 7.16 shows a smaller primary vortex while Fig. 7.17 shows a larger one, which is not unexpected.

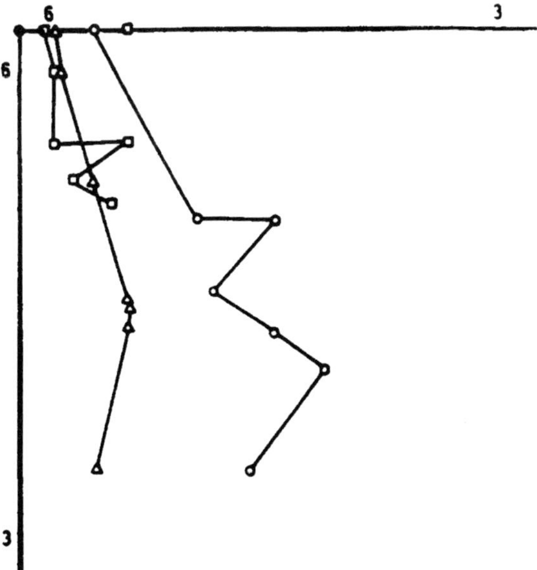

Figure 7.10 — Initial motion of P_1–P_4.

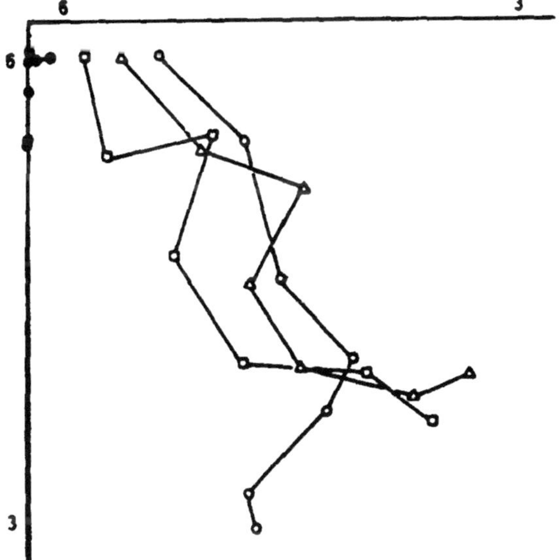

Figure 7.11 — Initial motion of P_{52}–P_{55}.

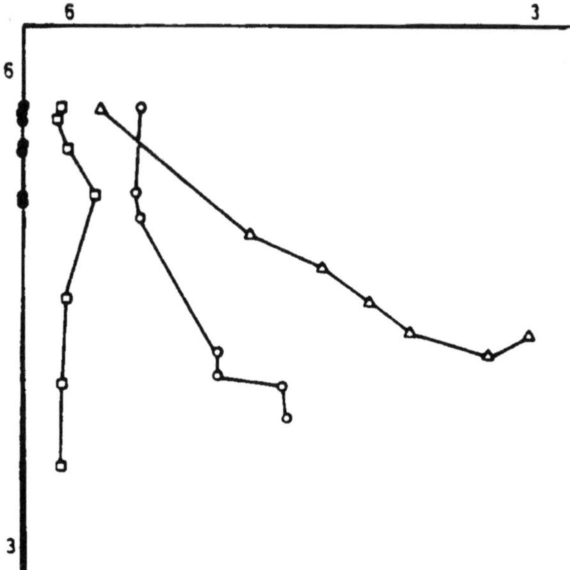

Figure 7.12 — Initial motion of P_{102}–P_{105}.

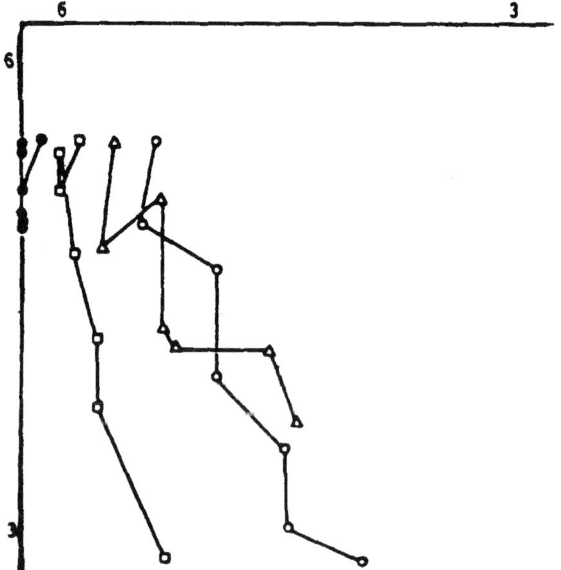

Figure 7.13 — Initial motion of P_{153}–P_{156}.

Figure 7.14 — Lower left corner secondary vortex.

Figure 7.15 — Lower right corner secondary vortex.

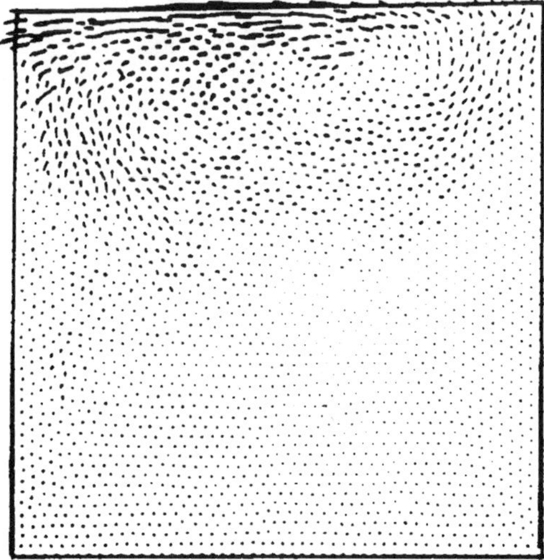

Figure 7.16 — Velocity field for $V = -7$.

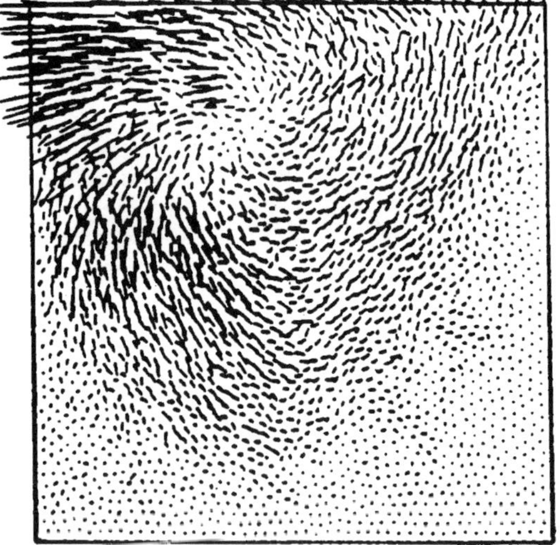

Figure 7.17 — Velocity field for $V = -13$ at t_{15000}.

8

Turbulent and Nonturbulent Vortices

8.1 Introduction

Since turbulence is the most commonly observed form of fluid behavior, interest has been exceptional in the modeling of related phenomena. Beginning with G. I. Taylor's seminal paper (Taylor (1921)), one school of study (Favre (1964)) has emphasized the statistical approach. A second major school of thought, following fundamental papers of Landau (1944), Hopf (1984), Lorenz (1963), and Ruelle and Takens (1971) uses Galerkin approximations to simplify the Navier-Stokes equations and bifurcation theory to analyze the resulting ordinary differential system (Barenblatt, Looss and Joseph (1983)). A more recent, computer-oriented approach is to solve the full Navier-Stokes equations numerically, assuming, of course, that these equations represent turbulent flow in some average sense (Markatos (1986)). In addition, there exists a variety of no less interesting, but less studied, approaches, as, for example, the thermodynamic model of Malkus (1960).

Unfortunately, important and realistic aspects of turbulent motions have defied inclusion in all the models described above (Favre (1964), Barenblatt, Looss and Joseph (1983), Markatos (1986), Saffman (1968)). Thus, for example, whereas homogeneous turbulence has received intensive theoretical study, it is not known to exist anywhere in Nature.

Our purpose in this chapter is to initiate a particle approach to the study of turbulence. In a natural way, turbulent behavior will be induced by conditions which permit the large, *repulsive* effects of intermolecular forces to prevail. The particle equations used will, in fact, constitute a primitive set of Navier-Stokes equations in the sense that they approximate a molecular formulation from which the continuous Navier-Stokes equations can be suitably derived (Hirschfelder, Curtiss and Bird (1954)).

Note, finally, that, for simplicity, attention will be restricted only to vortex motion.

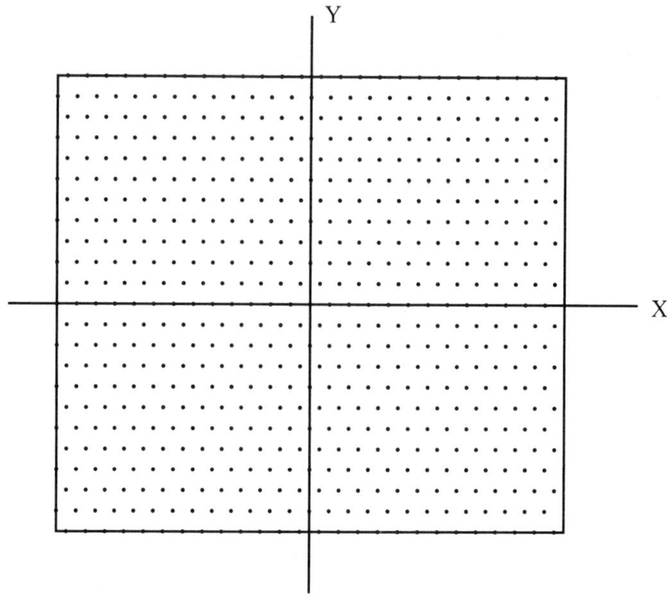

Figure 8.1.

8.2 Basic Definitions

Let us begin, rather specifically, by considering the 609 particle arrangement shown
in Fig. 8.1. (A more general approach follows in a natural way from the physical
assumptions to be made.)

The enclosed rectangle is bounded by the lines whose equations are $y = 9.526$,
$y = -9.526$, $x = 13.0$, $x = -13.0$, and the sides of the rectangle are called the
walls. The particles are arranged on and within the rectangle in such a fashion
that the distance between any particle and its immediate neighbors is unity. The
position (x_i, y_i) of each P_i is given by

$$x_1 = -12.5, \quad x_{27} = -13.0, \quad y_1 = -9.526, \quad y_{27} = 8.660,$$

$$x_{i+1} = 1.0 + x_i, \quad y_{i+1} = y_1; \quad i = 1, 2, \ldots, 25,$$

$$x_{i+1} = 1.0 + x_i, \quad y_{i+1} = y_{27}; \quad i = 27, 28, \ldots, 52,$$

$$x_i = x_{i-53}, \quad y_i = -1.732 + y_{i-53}; \quad i = 54, 55, \ldots, 609.$$

By construction, each particle in the interior of the rectangle has exactly six
immediate neighbors, each a unit away and on the vertices of a regular hexagon.

Next, as shown in Fig. 8.2, the seven particles P_{260}, P_{261}, P_{286}, P_{287}, P_{288},
P_{313}, P_{314} are assumed to form a solid hexagon, which will be allowed to move as

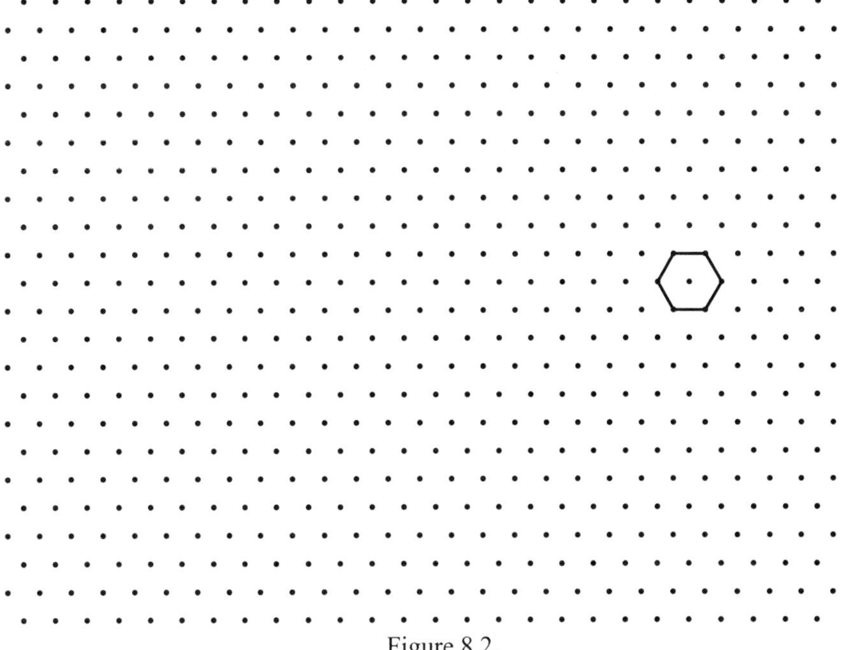

Figure 8.2.

a rigid body at a fixed velocity $\vec{V} = (VX, VY)$ per unit time. Our problem will be to describe the motion of the remaining particles, called the fluid particles, as the hexagon proceeds on its fixed, linear trajectory. For intuition, one can think in terms of a planar suspension of particles in a gas as the hexagon moves through it (van Dyke (1982)). Long-range forces will be considered to be negligible.

Now, as the hexagon moves, the fluid particles will experience displacement. Relative to these displacements, the following definitions are given for the fluid particles. Illustrative examples will be given later.

DEFINITION 8.1 For a given time step Δt, let P_i be located at $(x_{i,k}, y_{i,k})$ at time $t_k = k\Delta t$, $k = 0, 1, 2, \ldots$. Then P_i's instantaneous velocity $\vec{v}_{i,k}$ at t_k is defined by

$$\vec{v}_{i,k} = \left(\frac{x_{i,k} - x_{i,k-1}}{\Delta t}, \frac{y_{i,k} - y_{i,k-1}}{\Delta t} \right).$$

Definition 8.1 uses backward differences because these are consistent with whatever measurements an observer of the hexagon can have made at time t_k.

DEFINITION 8.2 For an integer $J > 1$, let P_i be at $(x_{i,k}, y_{i,k})$ at time t_k and at $(x_{i,k-J}, y_{i,k-J})$ at time $t_k - J\Delta t$. Then P_i's average velocity $\vec{v}_{i,k,J}$ at time t_k is defined by

$$\vec{v}_{i,k,J} = \left(\frac{x_{i,k} - x_{i,k-J}}{J\,\Delta t}, \; \frac{y_{i,k} - y_{i,k-J}}{J\,\Delta t} \right).$$

Definition 8.2 enables us to describe gross particle motions over extended periods by filtering the Brownian type local motions.

DEFINITION 8.3   If, relative to a particle P_i, the six particles closest to P_i have instantaneous velocities, relative to P_i, which are all either clockwise or counterclockwise, then the resulting seven-particle configuration is called an instantaneous vortex.

DEFINITION 8.4   If, relative to a particle P_i, the six particles closest to P_i have average velocities, relative to P_i, which are all either clockwise or counterclockwise, then the resulting seven-particle configuration is called an average vortex.

Definitions 8.3 and 8.4 prescribe the minimum size of the vortex that we will try to observe. Vortices of all magnitudes exist, from the microscopic to the macroscopic, but they can be recognized only within the limits of accuracy of one's observational instruments. Definitions 8.3 and 8.4 assume that we cannot observe vortices that have only five or fewer particles in clockwise or counterclockwise rotation relative to P_i.

DEFINITION 8.5   Relative to P_i and J, a vortex is said to be turbulent at time t_k if either of the following is valid: (a) It is an average vortex at t_k but not an instantaneous vortex at either t_k or t_{k-J}; (b) It is an instantaneous vortex at t_k but not an instantaneous vortex at t_{k-J}, not an average vortex at t_k, and not an instantaneous vortex at t_{k+J}.

Definition 8.5, part (a), allows local motions to be of Brownian type while average motions are not, whereas part (b) implies that for relatively small J the vortex appears and disappears quickly.

It should now be readily apparent that the choice of J in Definitions 8.2–8.5 is critical to an appropriate simulation of turbulence. A proper physical choice of J may, indeed, depend on position, time, $\vec{V}$, and/or other physical factors.

Note that a vortex that is not turbulent is called, quite naturally, a nonturbulent vortex.

The forces in our equations of motion are, again, for any two particles,

$$\vec{F}_{ij,k} = \left(\left[-\frac{G}{(r_{ij,k})^p} + \frac{H}{(r_{ij,k})^q} \right] \frac{\vec{r}_{ji,k}}{r_{ij,k}} \right), \qquad r_{ij,k} < D \qquad (8.1)$$

in which D is the distance of local interaction.

8.3 Examples

Throughout this section, set $\Delta t = 0.0001$ and, for reasons to be discussed later, set $J = 600$.

Example 8.1

Consider first the solid–fluid arrangement in Fig. 8.2. Fix $m_i = 1$, $i = 1, 2, 3, \ldots, 609$. Assume that the initial velocities of all the fluid particles are relatively negligible, that is, are much smaller than unity. For simplicity, then, set these to zero. Let the solid particles P_{260}, P_{261}, P_{286}, P_{287}, P_{288}, P_{313}, P_{314} each have $\vec{V} = (-7.5, -1.0)$, so that the hexagon moves linearly with this velocity through the fluid. This is implemented mechanically at each time step t_k by discarding the numerical calculations for the positions and velocities of the solid particles, decreasing their x and y coordinates by 0.00075 and 0.0001, respectively, and resetting their velocities to $(-7.5, -1.0)$. Let the parameters of local interaction be $G = 30$, $H = 50$, $p = 1$, $q = 3$, and $D = 1.5$, which allow the fluid particles to move relatively freely. Reflection from the walls is implemented by resetting particles that have moved out of the rectangle back to the nearest wall, in a perpendicular direction, with zero velocity.

The numerical results for the resulting fluid motion were studied every 600 time steps in the range t_{2400} to t_{12600} from the points of view of both instantaneous velocities and average velocities with $J = 600$. The value $J = 600$ was chosen, after several values of J were studied, because it proved to reveal vortex motion readily.

Figures 8.3–8.5 show the instantaneous velocity fields at the respective times t_{9600}, t_{10200}, t_{10800}. Figure 8.6 shows the average velocity field at t_{10200}, and Fig. 8.7 shows the average velocity field at t_{10200} relative to P_{258}, revealing that P_{258} is the center of a moving vortex. Particle P_{258} is marked crossed in Fig. 8.7. Moreover, though six points near P_{258} in both Figs. 8.3 and 8.4 are moving counterclockwise relative to P_{258}, the *closest* six points are not in such motion. Thus, the vortex shown in Fig. 8.7 is a turbulent vortex of the type (a) described in Definition 8.5. Moreover, Fig. 8.8, which shows the instantaneous velocity field at t_{10200} relative to P_{365}, reveals a second vortex which is an instantaneous vortex. As shown in Fig. 8.9,

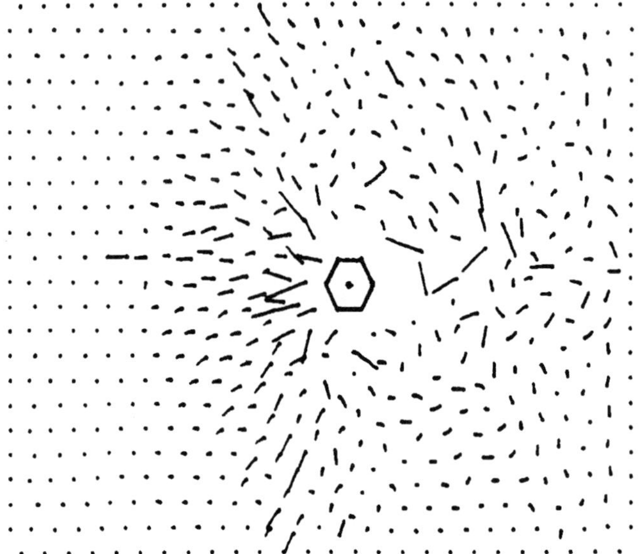

Figure 8.3 — Instantaneous velocity field at t_{9600}.

this vortex is *not* an average vortex at t_{10200}. Moreover, it is not an instantaneous vortex at either t_{9600} or t_{10800}, so it is a turbulent vortex of the type (b) described in Definition 8.5. Indeed, the vortex in Fig. 8.8 has resulted at t_{10200} from extensive counterclockwise motion just above it, and it then dissipates quickly with the counter-rotating vortex it adjoins.

Example 8.2
Example 8.1 was repeated with the single change $\vec{V} = (-5.0, -0.5)$, thus decreasing the speed and altering the direction of the hexagon. No vortices ever developed.

Example 8.3
Example 8.1 was repeated with the change $p = 2$, $q = 5$, thus making the fluid more cohesive. The velocity fields at t_{5400} and t_{6000} indicated the existence of an instantaneous frontal vortex between these time steps, which additional computation and printout confirmed. Figure 8.10 shows the instantaneous velocity field at t_{5700} relative to P_{362}, and reveals that P_{362} is the center of an instantaneous vortex. However, Fig. 8.11, six hundred time steps later, shows the average velocity field at t_{6300} and reveals that P_{362} is the center of an average vortex. Thus, the average vortex shown in Fig. 8.11 does not satisfy the conditions of Definition 8.5, so that it is a nonturbulent vortex.

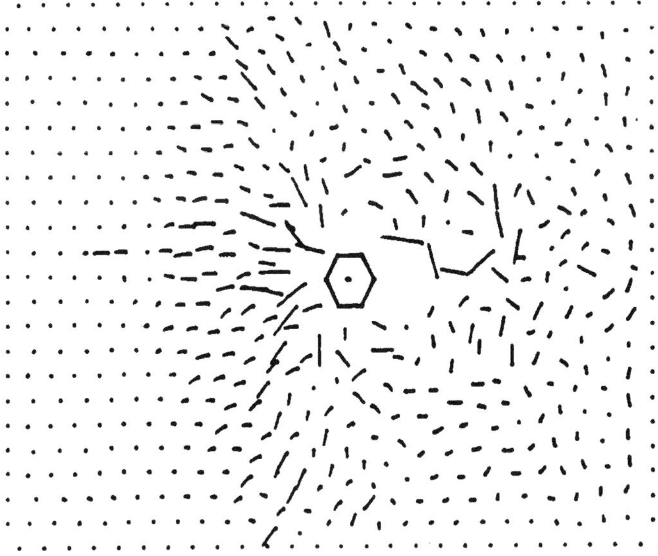

Figure 8.4 — Instantaneous velocity field at t_{10200}.

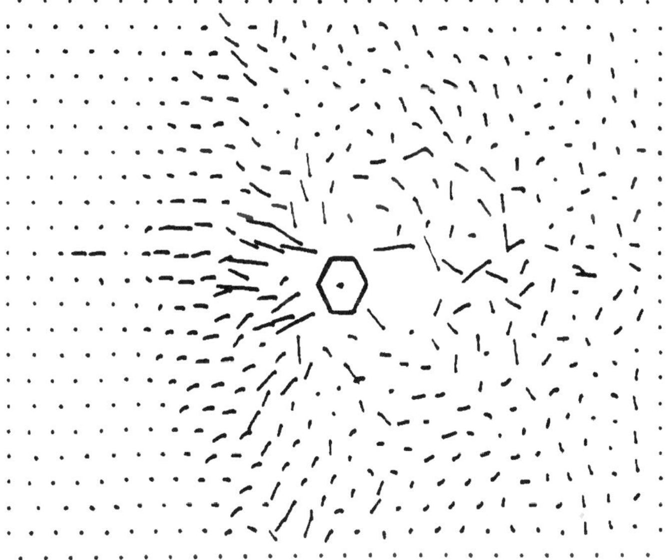

Figure 8.5 — Instantaneous velocity field at t_{10800}.

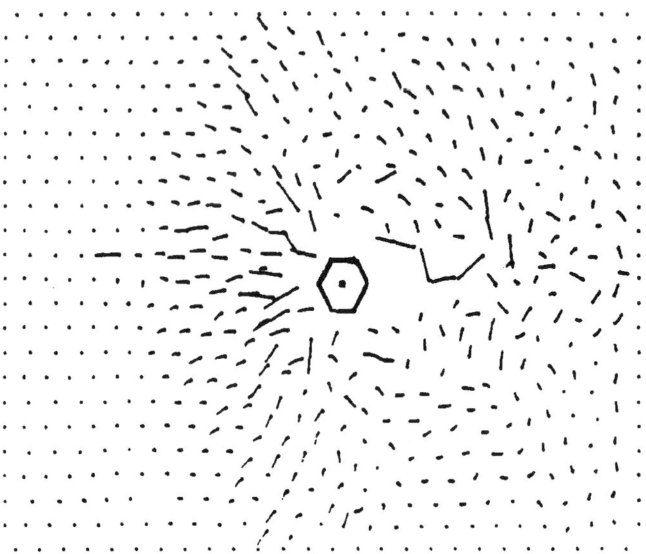

Figure 8.6 — Average velocity field at t_{10200}.

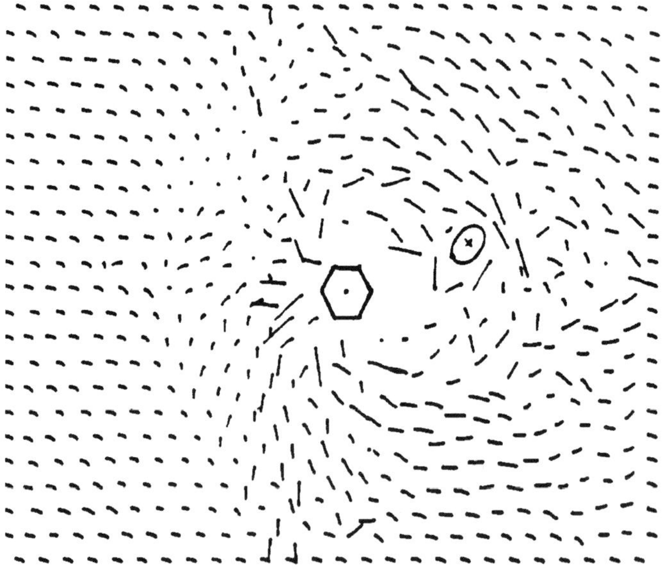

Figure 8.7 — Type (a) turbulent vortex relative to P_{258} at t_{10200}.

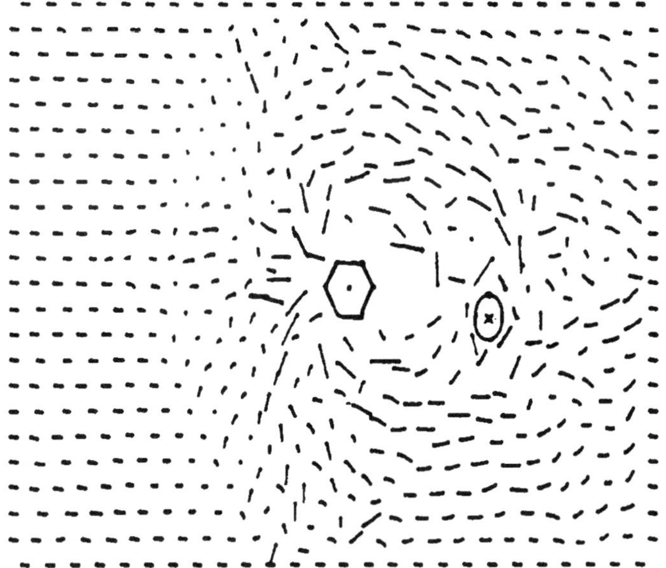

Figure 8.8 — Type (b) turbulent vortex relative to P_{365} at t_{10200}.

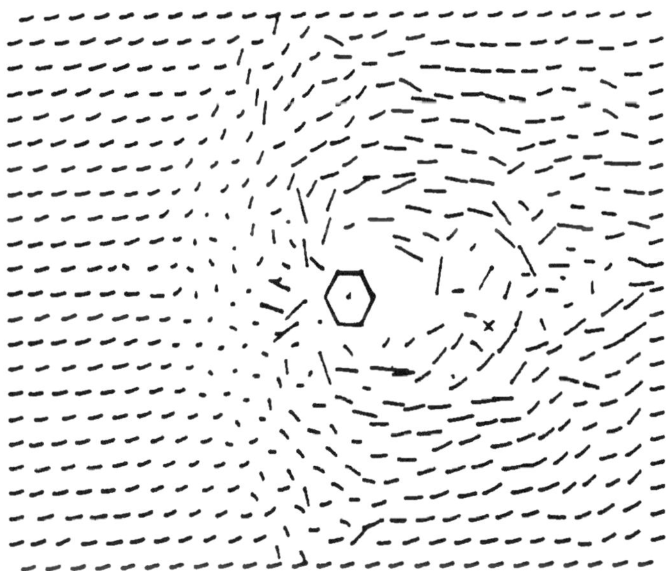

Figure 8.9 — Average velocity field relative to P_{365} at t_{10200}.

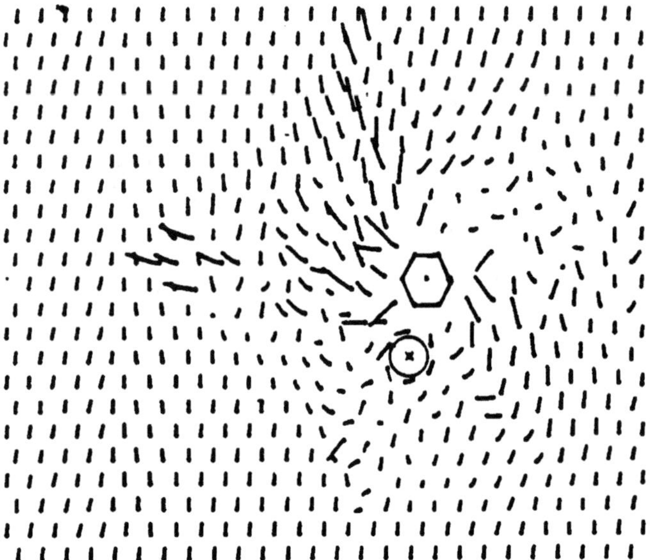

Figure 8.10 — Instantaneous velocity field relative to P_{362} at t_{5700}.

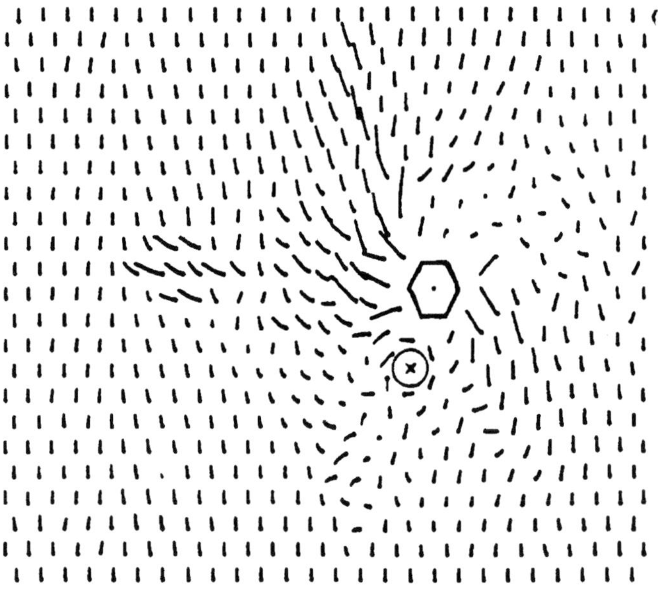

Figure 8.11 — Nonturbulent vortex relative to P_{362} at t_{6300}.

8.4 Remark

It is interesting to observe, from the examples of Sect. 8.3 and from additional examples, that nonturbulent vortices were relatively circular and the speeds of all six rotating particles, relative to the central particle, were about the same size. Turbulent vortices showed no such regularity.

9

Liquid Drop Formation, Fall, and Collision

9.1 Introduction

Liquid drop formation and interaction are important in fast kinetic, microwave, nucleation, and rainstorm studies (Peterson (1985), Simpson and Haller (1988)). In this chapter, we develop a generic model of such phenomena on the microdrop level. The present discussion is limited to two dimensions.

9.2 Drop Generation

Consider an open basin in the upper half of the xy plane that lies above the x axis and between the lines whose equations are $x = 25$, $x = -25$. Into this basin are placed 2500 particles $P_i, i = 1, 2, \ldots, 2500$, whose respective coordinates (x_i, y_i) are given by

$$x_1 = -25.0, \quad y_1 = 0.0, \quad x_{i+1} = 1.0 + x_i,$$

$$y_{i+1} = 0.0, \quad i = 1, 2, 3, \ldots, 50$$

$$x_{52} = -24.5, \quad y_{52} = 0.866, \quad x_{i+1} = 1.0 + x_i,$$

$$y_{i+1} = 0.866, \quad i = 52, 53, \ldots, 100$$

$$x_i = x_{i-101}, \quad y_i = 1.732 + y_{i-101}, \quad i = 102, 103, \ldots, 2500.$$

A velocity is prescribed for each particle in an arbitrary fashion as follows. The velocity $\vec{v}_i = (v_{i,x}, v_{i,y})$ of P_i has zero components except for $v_{i,x} = 0.01, i = 1, 2, \ldots, 200; v_{i,y} = -0.01, i = 700, 701, \ldots, 900; v_{i,x} = -0.01, i = 1100, 1101, \ldots, 1300; v_{i,y} = 0.01, i = 1450, 1451, \ldots, 1650; v_{iy} = 0.005, i = 2000, 2001,$

..., 2450. Next, the 2500 particles are allowed to interact dynamically as follows. Each particle is acted upon by gravity and, locally only, by a classical molecular type force whose magnitude F is given by

$$F = -\frac{25}{r^3} + \frac{10}{r^5} \tag{9.1}$$

The distance of local interaction D is taken to be 5. The gravity constant g is scaled to 9.8. To keep the fluid within the basin, particles are reflected symmetrically from the boundaries but with a velocity damping factor $\delta = 0.9$. Assuming unit mass for each particle, the Newtonian dynamical equations of motion are solved numerically by the leap frog formulas at $t_k = (0.0002)k$, $k = 1, 2, \ldots, 77000$, after which all particles with y coordinates greater than 8 are dropped from consideration. There remain 1454 particles. The basin boundaries are now discarded, and the particles are inverted by symmetrical reflection about the x axis. In addition, 201 more particles are generated to simulate a solid ceiling by setting these at the points $(-25.0, 0.25)$, $(-24.75, 0.25)$, $(-24.50, 0.25)$, $\ldots$, $(25.0, 0.25)$. Thus a system of 1655 particles has been generated, of which 1454 are fluid particles, while 201 are fixed, solid particles. The resulting system is shown in Fig. 9.1(a).

To create a fluid drop that is attached to the ceiling, we proceed as follows. Fluid particles are allowed to interact dynamically with both fluid and solid particles, but motion is allowed only to the fluid particles. The dynamical parameters are now reset so that $g = 0.98$; the force for fluid–fluid interaction is taken to have magnitude

$$F_1 = -\frac{200}{r^3} + \frac{80}{r^5};$$

and the force for fluid–solid interaction is taken to be

$$F_2 = -\frac{250}{r^3} + \frac{100}{r^5}.$$

All other parameter choices remain the same. For display purposes, the counter k is reset to zero. The numerical solution is shown in Figs. 9.1(b)–(d) at $k = 1000$, 6000 and 11,000. At each of these times, the system kinetic energy was relatively high, so at $k = 11,000$ and every 1000 steps thereafter until $k = 16,000$ the fluid particle velocities are damped by the factor 0.2. The result is shown in Fig. 9.1(e), in which 56 particles that have succumbed to gravity have been eliminated. The number of particles remaining in the drop is 1398, and the drop shape shown is in complete agreement with experimental results.

9.3 Drop Fall

Beginning now with the drop configuration in Fig. 9.1(e), the simulation was continued as follows. All damping was removed, and the single parameter g was

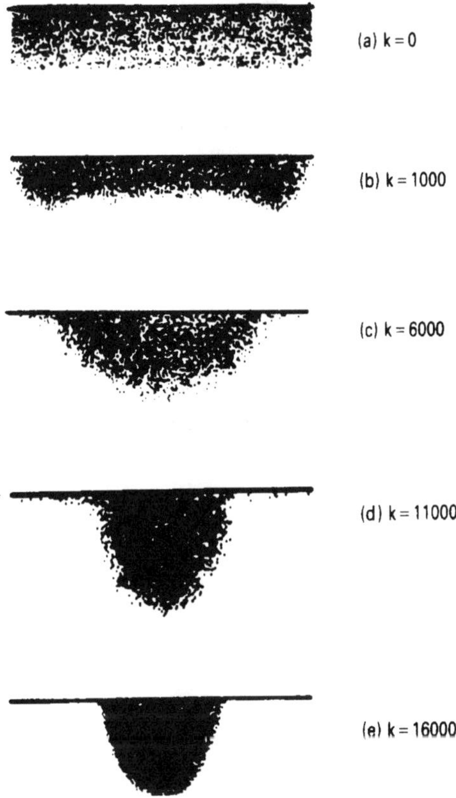

Figure 9.1 — Drop formation

reset from 0.98 to 98.0. For display purposes the counter k was again reset to zero. Figure 9.2 shows the initial fall of the drop for $k = 1000, 2000, \ldots , 5000$. The figure shows fluid contraction along the ceiling interface, the development of a fluid neck, elongation of both the neck and the main body of the drop, a breaking of the neck, and the formation of a falling drop.

As the drop falls, its speed increases, so its shape becomes dependent on the medium in which it falls. Thus, after $k = 5000$ the effect of this medium was introduced as follows. The region of the xy plane between the lines with equations $x = -25$, $x = 25$ was subdivided into 50 nonoverlapping vertical strips, each of unit width. As the drop continued to fall, the lowest particle in each strip was damped every 500 steps by having its vertical component of velocity multiplied by the factor d. Three cases were considered: $d = 1.0, d = 0.9$ and $d = 0.6$. The results are shown in Fig. 9.3 for $k = 6000, 7000, 8000$, and 9,000. The leftmost part of the figure shows fall in a vacuum, that is, for $d = 1.0$. The rightmost part of the figure shows the fall for $d = 0.6$. The figures reveal a growing horizontal oblateness

Figure 9.2 — Drop fall.

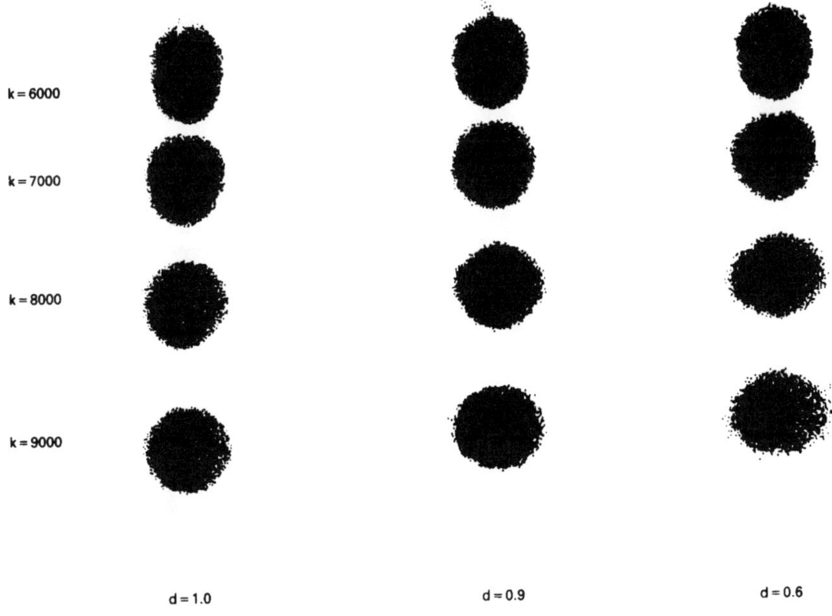

Figure 9.3 — Continued fall.

due to an increase of pressure from below. After $k = 9000$, computations reveal relatively small system oscillations, which probably result from the inital stretching of the drop, molecular interaction, and when present, pressure from below. Simulation with $d = 0.3$ resulted in complete drop disintegration.

9.4 Drop Collision

To study drop collision, the relatively circular drop configuration on the bottom of the leftmost column of Fig. 9.3 was isolated, and a symmetric image, with the same particle velocities was generated about the y axis, as shown in Fig. 9.4(a). To study modes of collision, the image system to the right of the y axis had each of its particle velocities decreased by $\Delta v_x = -v_1$, $\Delta v_y = -v_2$, where the actual velocity components of any particle are relatively small with respect to v_1 and/or v_2. For simplicity, g was reset to zero, and all ceiling particles and stray particles were eliminated, so the entire system shown in Fig. 9.4(a) consists of $2(1367) = 2734$ particles.

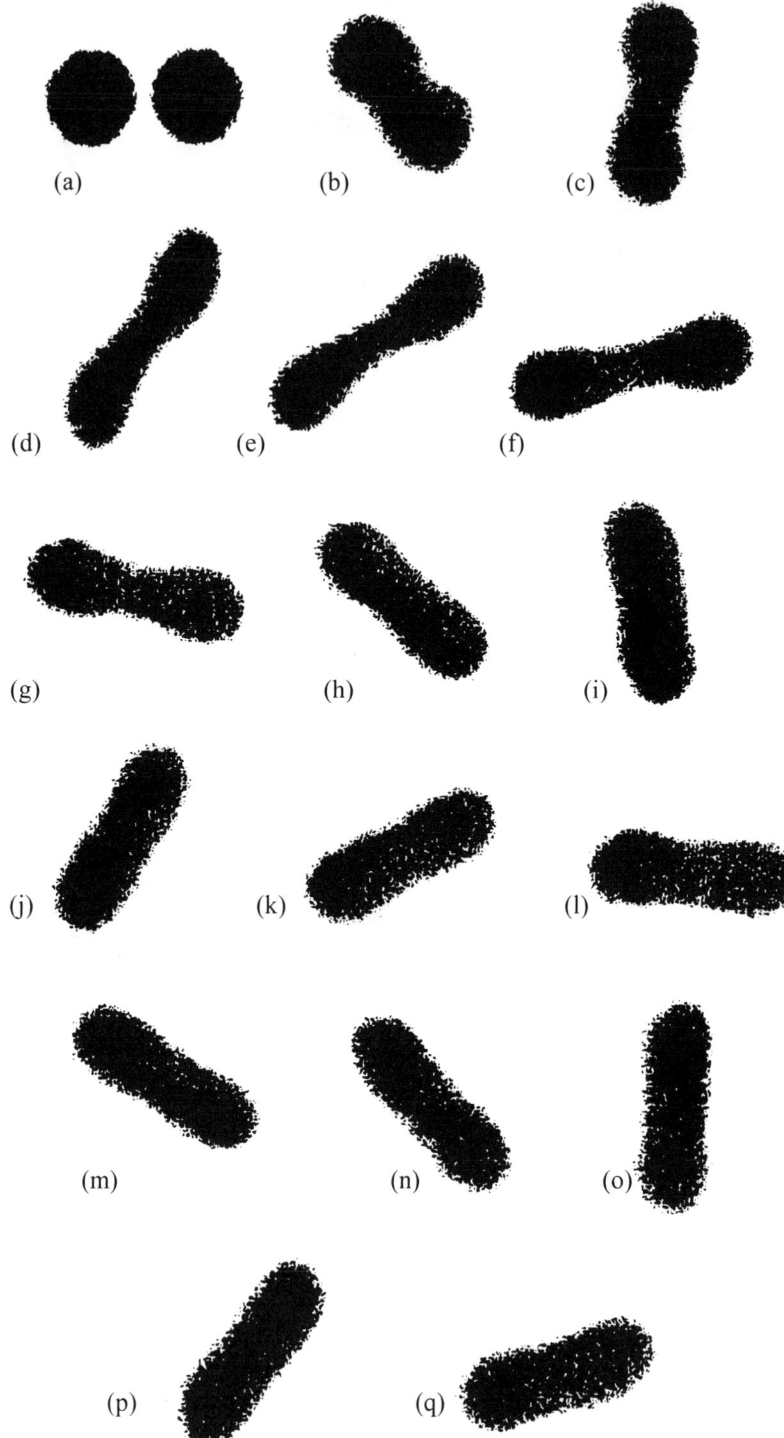

Figure 9.4 — Drop collision with $V_1 = 20$, $V_2 = 30$.

For $v_1 = 20$, $v_2 = 30$, Fig. 9.4 shows the oscillatory mode for every 2000 time steps through $k = 32,000$. This "rotating dumbell" mode is well known to meteorologists and chemists (Simpson and Haller (1988)).

For $v_1 = 10$, $v_2 = 0$, Fig. 9.5 shows, for every 1000 time steps, the resulting simple oscillatory mode in which vertical and horizontal oblateness alternate in a consecutive fashion.

Finally, for $v_1 = 50$, $v_2 = 0$, Fig. 9.6 shows, for every 1000 time steps a collision in which no vibrational mode results because excessive momentum has resulted in the creation of two new drops, each of which shares approximately half of the particles of the original drops.

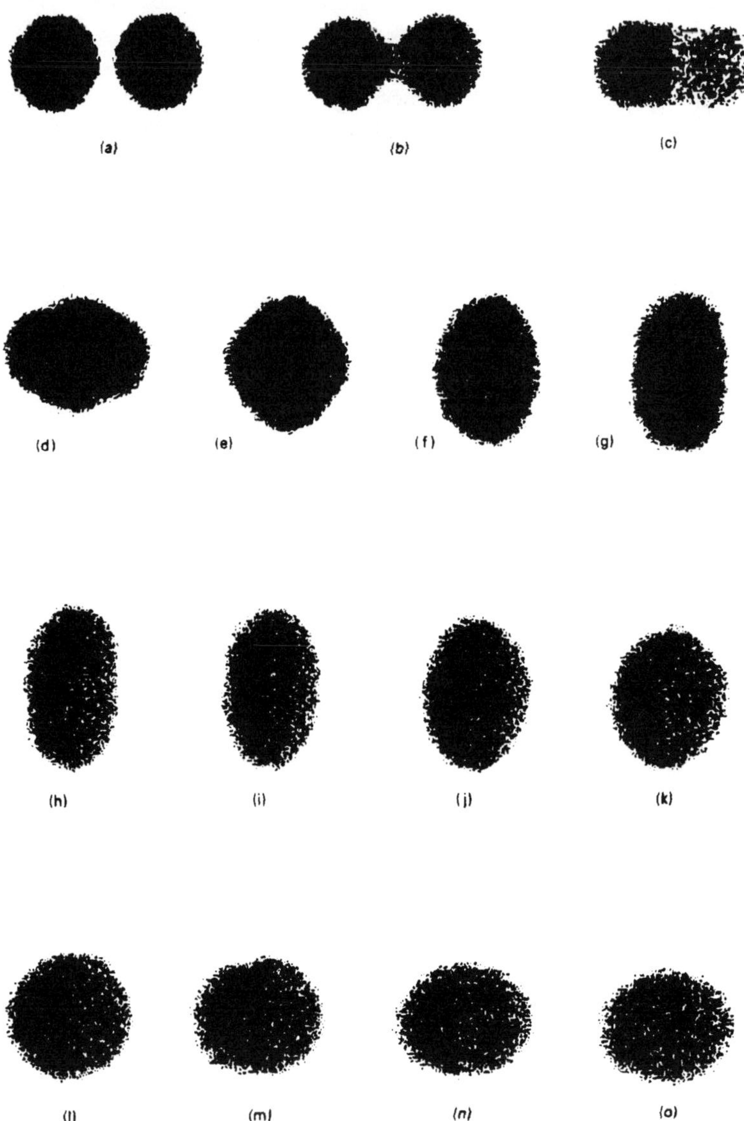

Figure 9.5 — Drop collision with $V_1 = 10$, $V_2 = 0$.

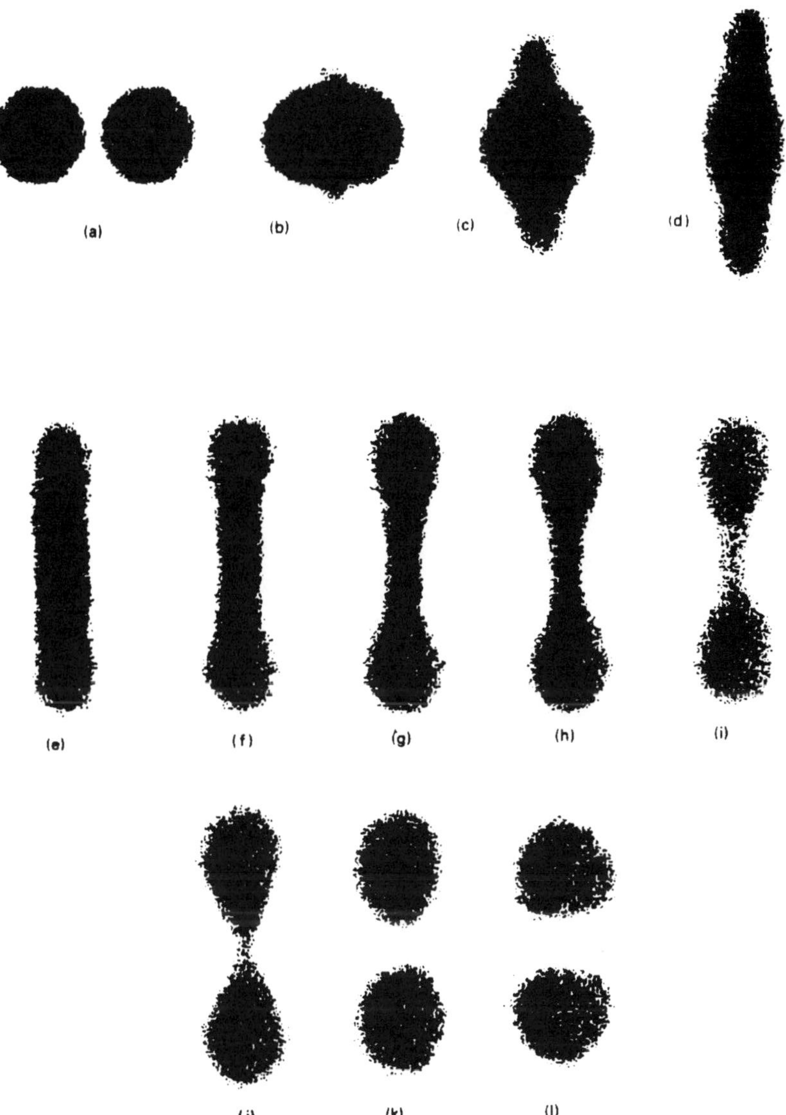

Figure 9.6 — Drop collision with $V_1 = 50$, $V_2 = 0$.

10

Conservative Motion of Tops and Gyroscopes

10.1 Introduction

Rigid body motion is of fundamental interest in mathematics, science and engineering. In this chapter we will introduce a simplistic approach to this area of study in the spirit of modern molecular mechanics. We will consider first a discrete tetrahedral body and simulate its motion when it spins like a top whose contact point with the XY plane is allowed to move in the plane. The approach will not require the use of special coordinates, Cayley–Klein parameters, tensors, dyadics, or related concepts (Goldstein (1980)). All that will be required is Newtonian mechanics in three-dimensional XYZ space.

10.2 A Discrete, Rigid Tetrahedral Top

Consider as shown in Fig. 10.1 a regular tetrahedron with vertices $P_i(x_i, y_i, z_i)$, $i = 1, 2, 3, 4$, and edge length R. For convenience set

$$(x_1, y_1, z_1) = (0, 0, R\sqrt{6}/3), \quad (x_2, y_2, z_2) = (0, R\sqrt{3}/3, 0)$$

$$(x_3, y_3, z_3) = \left(\frac{1}{2}R, -R\sqrt{3}/6, 0\right), \quad (x_4, y_4, z_4) = \left(-\frac{1}{2}R, -R\sqrt{3}/6, 0\right).$$

The geometric center of triangle $P_2 P_3 P_4$ is $(0, 0, 0)$ and the geometric center of the tetrahedron is $(0, 0, R\sqrt{6}/12)$.

In order to create a top, let us first invert the tetrahedron in Fig. 10.1 to the position shown in Fig. 10.2, so that

$$(x_1, y_1, z_1) = (0, 0, 0), \quad (x_2, y_2, z_2) = (0, R\sqrt{3}/3, R\sqrt{6}/3)$$

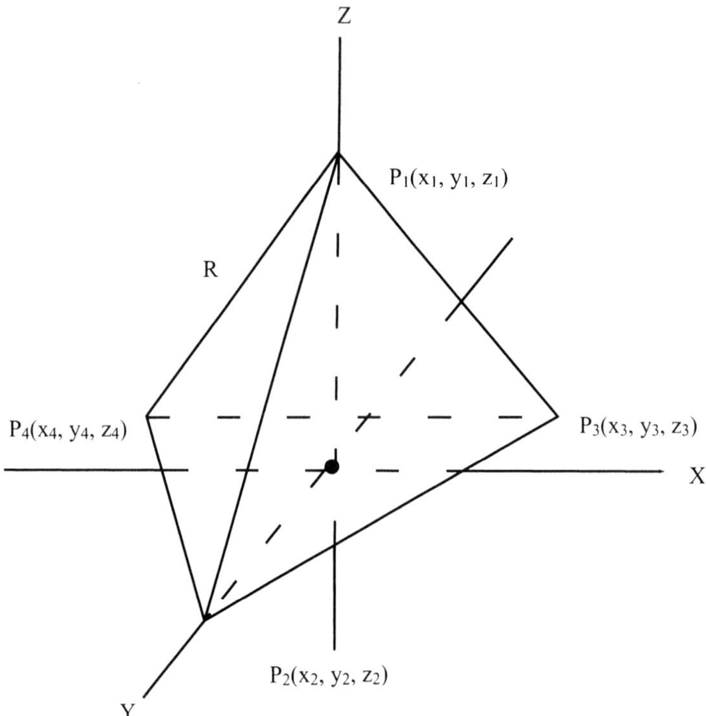

Figure 10.1.

$$(x_3, y_3, z_3) = \left(\frac{1}{2}R, -R\sqrt{3}/6, R\sqrt{6}/3 \right)$$

$$(x_4, y_4, z_4) = \left(-\frac{1}{2}R, -R\sqrt{3}/6, R\sqrt{6}/3 \right).$$

The geometric center of triangle $P_2 P_3 P_4$ is now $(0, 0, R\sqrt{6}/3)$, while the geometric center $\bar{P} = (\bar{x}, \bar{y}, \bar{z})$ of the inverted tetrahedron is $(0, 0, R\sqrt{6}/4)$.

Next, let us set P_2, P_3, P_4 in rotation in the plane $z = R\sqrt{6}/3$. As shown in Fig. 10.3, we let the velocity of each particle be perpendicular to the line joining that particle to the center of triangle $P_2 P_3 P_4$. Let P_2, P_3, P_4 each have the same speed V. Thus, we take the velocities $\vec{v}_i = (v_{ix}, v_{iy}, v_{iz}), i = 1, 2, 3, 4$, of P_1, P_2, P_3, P_4 to be

$$\vec{v}_1 = (0, 0, 0), \quad \vec{v}_2 = (V, 0, 0),$$

$$\vec{v}_3 = \left(-\frac{1}{2}V, -V\sqrt{3}/2, 0 \right), \quad \vec{v}_4 = \left(-\frac{1}{2}V, V\sqrt{3}/2, 0 \right).$$

Finally, we want the rotating tetrahedron to be tilted, initially, relative to the Z axis, so that we assume the line joining P_1 to $\bar{P}$ forms an angle α relative to the Z

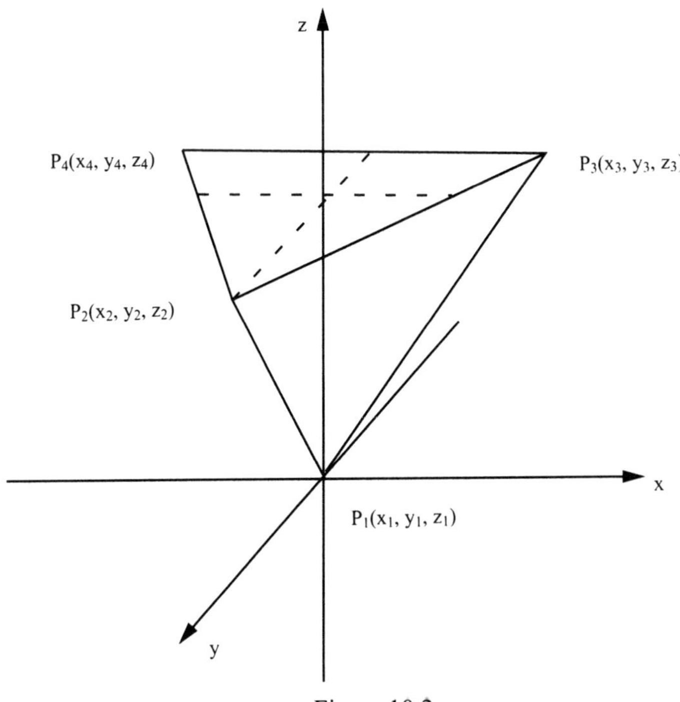

Figure 10.2.

axis. As shown in Fig. 10.4, this will be done by rotating the XZ plane through an angle α. Thus, the new positions (x_i', y_i', z_i') and the new velocities $(v_{ix'}, v_{iy'}, v_{iz'})$, $i = 1, 2, 3, 4$, satisfy

$$x_i' = x_i \cos\alpha + z_i \sin\alpha, \quad y_i' = y_i, \quad z_i' = -x_i \sin\alpha + z_i \cos\alpha,$$

$$v_{ix'} = v_{ix} \cos\alpha + v_{iz} \sin\alpha, \quad v_{iy'} = v_{iy}, \quad v_{iz'} = -v_{ix} \sin\alpha + v_{iz} \cos\alpha.$$

Thus, once the parameters R, V, and α are given, all initial data for a tilted, rotating tetrahedron are determined. Moreover, because of the top's symmetry, its motion will be characterized completely by the motions of P_1 and $\bar{P}$.

10.3 Dynamical Equations

The motion of our rotating top is now treated as a four-body problem. At any time t, let P_i, $i = 1, 2, 3, 4$, be located at $\vec{r} = (x_i, y_i, z_i)$, have velocity $\vec{v}_i = (\dot{x}_i, \dot{y}_i, \dot{z}_i) = (v_{ix}, v_{iy}, v_{iz})$, and have acceleration $\vec{a}_i = (\ddot{x}_i, \ddot{y}_i, \ddot{z}_i) = (\dot{v}_{ix}, \dot{v}_{iy}, \dot{v}_{iz})$. Let the mass of each P_i be m_i. For $i \neq j$, let $\vec{r}_{ij}$ be the vector from P_i to P_j and let r_{ij} be the magnitude of $\vec{r}_{ij}$, $i = 1, 2, 3, 4$; $j = 1, 2, 3, 4$; $i \neq j$. Let $\phi = \phi(r_{ij})$ be a potential

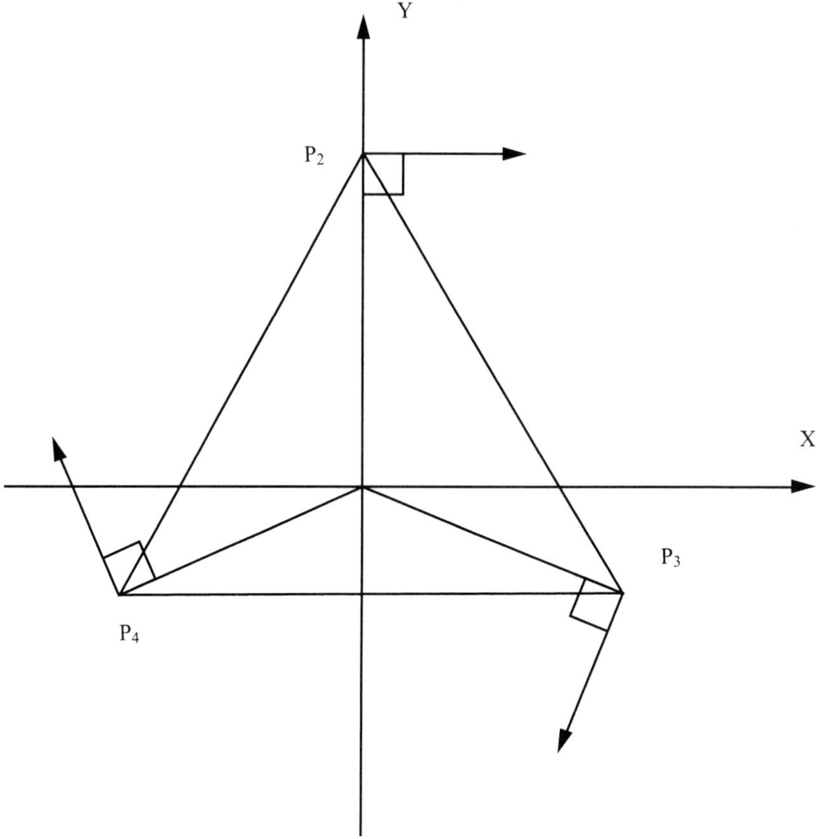

Figure 10.3.

function defined by the pair P_i, P_j, $i \neq j$. Then, for $i = 1, 2, 3, 4$, the Newtonian dynamical equations for the motion of the particles are the second-order differential equations

$$m_i a_{ix} = -\frac{\partial \phi}{\partial r_{ij}} \frac{x_i - x_j}{r_{ij}} - \frac{\partial \phi}{\partial r_{ik}} \frac{x_i - x_k}{r_{ik}} - \frac{\partial \phi}{\partial r_{im}} \frac{x_i - x_m}{r_{im}} \qquad (10.1)$$

$$m_i a_{iy} = -\frac{\partial \phi}{\partial r_{ij}} \frac{y_i - y_j}{r_{ij}} - \frac{\partial \phi}{\partial r_{ik}} \frac{y_i - y_k}{r_{ik}} - \frac{\partial \phi}{\partial r_{im}} \frac{y_i - y_m}{r_{im}} \qquad (10.2)$$

$$m_i a_{iz} = -\frac{\partial \phi}{\partial r_{ij}} \frac{z_i - z_j}{r_{ij}} - \frac{\partial \phi}{\partial r_{ik}} \frac{z_i - z_k}{r_{ik}} - \frac{\partial \phi}{\partial r_{im}} \frac{z_i - z_m}{r_{im}} - g_i \qquad (10.3)$$

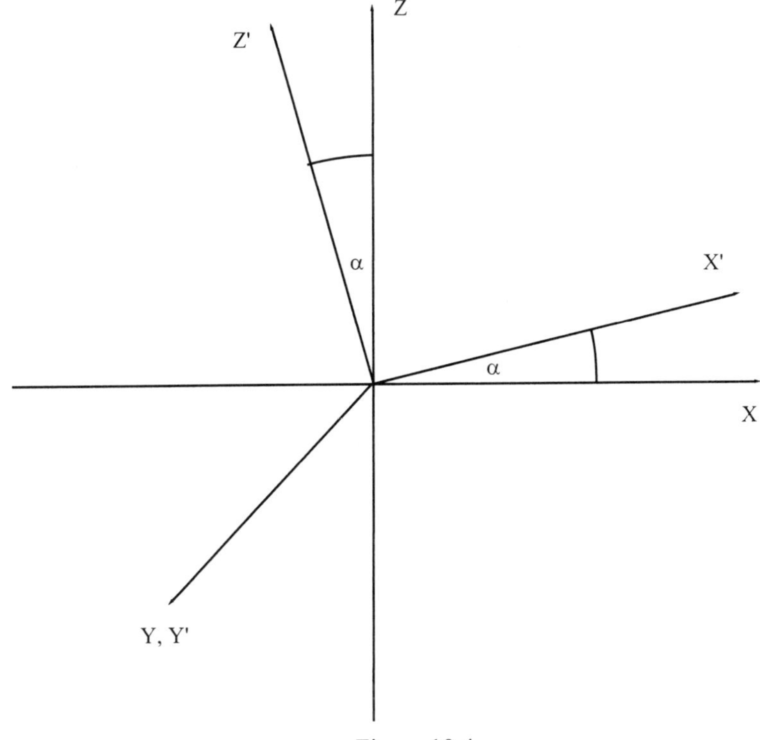

Figure 10.4.

in which $i = 1$ implies $j = 2, k = 3, m = 4$; $i = 2$ implies $j = 1, k = 3, m = 4$; $i = 3$ implies $j = 1, k = 2, m = 4$; $i = 4$ implies $j = 1, k = 2, m = 3$. Moreover, we choose $g_1 = 0$, $g_2 = g_3 = g_4 > 0$, so that gravity acts only on P_2, P_3, and P_4.

Equations (10.1)–(10.3) are fully conservative, that is, they conserve system energy, linear momentum and angular momentum.

10.4 Numerical Method

In order to solve system (10.1)–(10.3) in a fashion which conserves exactly the same energy, linear momentum and angular momentum, we first rewrite (10.1)–(10.3) as the following equivalent first-order system:

$$\frac{dx_i}{dt} = v_{ix} \tag{10.4}$$

$$\frac{dy_i}{dt} = v_{iy} \tag{10.5}$$

$$\frac{dz_i}{dt} = v_{iz} \tag{10.6}$$

$$m_i \frac{dv_{ix}}{dt} = -\frac{\partial \phi}{\partial r_{ij}} \frac{x_i - x_j}{r_{ij}} - \frac{\partial \phi}{\partial r_{ik}} \frac{x_i - x_k}{r_{ik}} - \frac{\partial \phi}{\partial r_{im}} \frac{x_i - x_m}{r_{im}} \tag{10.7}$$

$$m_i \frac{dv_{iy}}{dt} = -\frac{\partial \phi}{\partial r_{ij}} \frac{y_i - y_j}{r_{ij}} - \frac{\partial \phi}{\partial r_{ik}} \frac{y_i - y_k}{r_{ik}} - \frac{\partial \phi}{\partial r_{im}} \frac{y_i - y_m}{r_{im}} \tag{10.8}$$

$$m_i \frac{dv_{iz}}{dt} = -\frac{\partial \phi}{\partial r_{ij}} \frac{z_i - z_j}{r_{ij}} - \frac{\partial \phi}{\partial r_{ik}} \frac{z_i - z_k}{r_{ik}} - \frac{\partial \phi}{\partial r_{im}} \frac{z_i - z_m}{r_{im}} - g_i. \tag{10.9}$$

We now choose difference equation approximations for (10.4)–(10.9). For a fixed time step Δt, let $t_n = n\Delta t$, $n = 0, 1, 2, 3, \ldots$. At t_n let P_i be at $\vec{r}_{i,n} = (x_{i,n}, y_{i,n}, z_{i,n})$ with velocity $\vec{v}_{i,n} = (v_{i,x,n}, v_{i,y,n}, v_{i,z,n})$. Then, in accordance with Sect. 2.3, equations (10.4)–(10.9) will be approximated by

$$\frac{x_{i,n+1} - x_{i,n}}{\Delta t} = \frac{v_{i,x,n+1} + v_{i,x,n}}{2} \tag{10.10}$$

$$\frac{y_{i,n+1} - y_{i,n}}{\Delta t} = \frac{v_{i,y,n+1} + v_{i,y,n}}{2} \tag{10.11}$$

$$\frac{z_{i,n+1} - z_{i,n}}{\Delta t} = \frac{v_{i,z,n+1} + v_{i,z,n}}{2} \tag{10.12}$$

$$m_i \frac{v_{i,x,n+1} - v_{i,x,n}}{\Delta t} = -\frac{\phi(r_{ij,n+1}) - \phi(r_{ij,n})}{r_{ij,n+1} - r_{ij,n}} \cdot \frac{x_{i,n+1} + x_{i,n} - x_{j,n+1} - x_{j,n}}{r_{ij,n+1} + r_{ij,n}}$$

$$- \frac{\phi(r_{ik,n+1}) - \phi(r_{ik,n})}{r_{ik,n+1} - r_{ik,n}} \cdot \frac{x_{i,n+1} + x_{i,n} - x_{k,n+1} - x_{k,n}}{r_{ik,n+1} + r_{ik,n}}$$

$$- \frac{\phi(r_{im,n+1}) - \phi(r_{im,n})}{r_{im,n+1} - r_{im,n}} \cdot \frac{x_{i,n+1} + x_{i,n} - x_{m,n+1} - x_{m,n}}{r_{im,n+1} + r_{im,n}} \tag{10.13}$$

$$m_i \frac{v_{i,y,n+1} - v_{i,y,n}}{\Delta t} = -\frac{\phi(r_{ij,n+1}) - \phi(r_{ij,n})}{r_{ij,n+1} - r_{ij,n}} \cdot \frac{y_{i,n+1} + y_{i,n} - y_{j,n+1} - y_{j,n}}{r_{ij,n+1} + r_{ij,n}}$$

$$- \frac{\phi(r_{ik,n+1}) - \phi(r_{ik,n})}{r_{ik,n+1} - r_{ik,n}} \cdot \frac{y_{i,n+1} + y_{i,n} - y_{k,n+1} - y_{k,n}}{r_{ik,n+1} + r_{ik,n}}$$

$$- \frac{\phi(r_{im,n+1}) - \phi(r_{im,n})}{r_{im,n+1} - r_{im,n}} \cdot \frac{y_{i,n+1} + y_{i,n} - y_{m,n+1} - y_{m,n}}{r_{im,n+1} + r_{im,n}} \tag{10.14}$$

$$m_i \frac{v_{i,z,n+1} - v_{i,z,n}}{\Delta t} = -\frac{\phi(r_{ij,n+1}) - \phi(r_{ij,n})}{r_{ij,n+1} - r_{ij,n}} \cdot \frac{z_{i,n+1} + z_{i,n} - z_{j,n+1} - z_{j,n}}{r_{ij,n+1} + r_{ij,n}}$$

$$-\frac{\phi(r_{ik,n+1}) - \phi(r_{ik,n})}{r_{ik,n+1} - r_{ik,n}} \cdot \frac{z_{i,n+1} + z_{i,n} - z_{k,n+1} - z_{k,n}}{r_{ik,n+1} + r_{ik,n}}$$

$$-\frac{\phi(r_{im,n+1}) - \phi(r_{im,n})}{r_{im,n+1} - r_{im,n}} \cdot \frac{z_{i,n+1} + z_{i,n} - z_{m,n+1} - z_{m,n}}{r_{im,n+1} + r_{im,n}} - g_i. \qquad (10.15)$$

Difference equations (10.10)–(10.15) are consistent with differential equations (10.4)–(10.9) and conserve exactly the same system invariants as do (10.1)–(10.3). System (10.10)–(10.15), for each of $n = 0, 1, 2, \ldots$, consists of 24 equations in the unknowns $x_{i,n+1}, y_{i,n+1}, z_{i,n+1}, v_{i,x,n+1}, v_{i,y,n+1}, v_{i,z,n+1}, i = 1, 2, 3, 4$; and in the knowns $x_{i,n}, y_{i,n}, z_{i,n}, v_{i,x,n}, v_{i,y,n}, v_{i,z,n}$.

The FORTRAN program in Appendix A4 modifies readily for the motion of a top.

10.5 Examples

In considering examples, we must first choose a potential function ϕ. We do this in cgs units and in such a fashion that we ensure that the tetrahedron is rigid. To accomplish this, we introduce the following classical, molecular type function:

$$\phi = A\left[-\frac{1}{r_{ij}^3} + \frac{1}{r_{ij}^5}\right], \qquad A > 0, \qquad (10.16)$$

in which A is sufficiently large to ensure rigidity. The choice of exponents in (10.16) prevents the explosive behavior characteristic of real molecules, yet allows for attractive and repulsive interaction.

From (10.16), it follows that the magnitude F of the force $\vec{F}$ determined by ϕ satisfies

$$F = A\left[-\frac{3}{r_{ij}^4} + \frac{5}{r_{ij}^6}\right].$$

Thus, $F(\bar{r}) = 0$ provided $\bar{r} = (5/3)^{\frac{1}{2}} \sim 1.290994449$.

We now choose the tetrahedral edge length R to be

$$R = 1.290994449. \qquad (10.17)$$

For this value of R, the force between any two of the particles is zero, so that the tetrahedron is physically stable. In the examples to be described, the parameters A, g and m_i are scaled for computational convenience to be $A = 10^6$, $g = 0.980$, $m_i = 1, i = 1, 2, 3, 4$, unless otherwise specified.

The time step Δt is chosen to be $\Delta t = 10^{-5}$. The positions of P_1 and $\bar{P}$ are recorded every 5000 time steps through 11,000,000 time steps. Thus, 2200 points are available for each trajectory to be described. In the figures to be given, units on the Z axis are often rescaled to accentuate the character of the resulting trajectory. In all the examples, the distance between any of P_1, P_2, P_3, P_4 is always 1.291. If at any time any one of z_2, z_3, z_4 is zero, the calculations are stopped and it is concluded that the top has fallen and ceased its motion.

To assure that the motion of P_1 is in the XY plane, we assume throughout that $v_{1,z,n} = 0$, $n = 0, 1, 2, 3, \ldots$.

Example 10.1
Set $V = 4$, $\alpha = 15°$. Figure 10.5 shows the cusped path of P_1 in the XY plane using the first 350 points of its trajectory and yields just over one complete cycle. There are 6^+ cycles in the 2200 point trajectory shown in Fig. 10.6 which has been enlarged for clarity. The point $\bar{P}$ for the entire 2200 point trajectory oscillates on the line $(0.2046144211, 0, z)$, with $\bar{z}$ rising and falling in the range $0.754 < \bar{z} < 0.764$. Thus, the center point of the trajectories shown in Figs. 10.5 and 10.6 in the XY plane is $(0.2046144211, 0)$.

Example 10.2
Set $V = 8$, $\alpha = 15°$. Figure 10.7 shows the cusped trajectory of P_1 in the XY plane using the first 750 points of its trajectory to yield just over one complete cycle. There are 3^+ cycles in the 2200 point graph shown in Fig. 10.8 which has been enlarged for clarity. The point $\bar{P}$ for the entire 2200 point trajectory lies on the line $(0.2046144211, 0, z)$, that is, the same line as in Example 1, but with $\bar{z}$

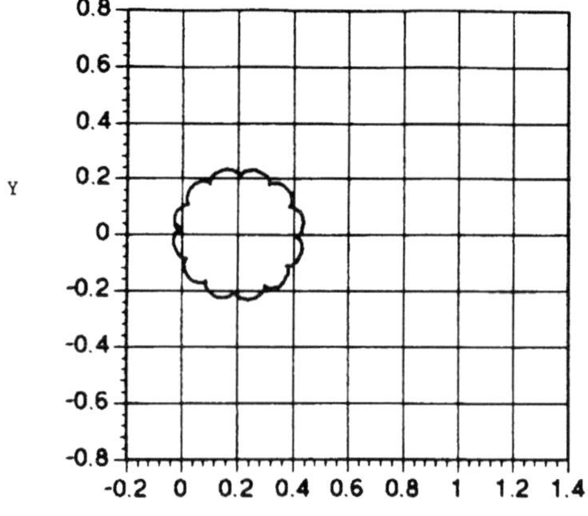

Figure 10.5.

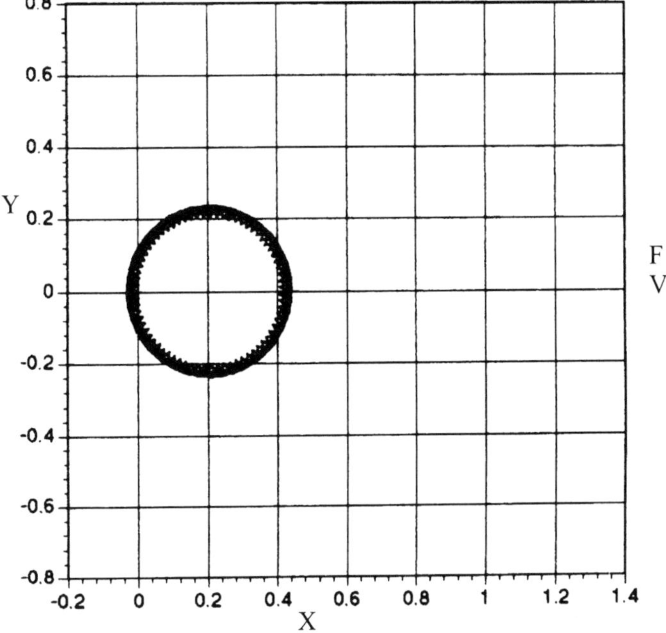

Figure 10.6.

rising and falling in the range $0.762 < \bar{z} < 0.764$. Thus, the center point of the trajectories shown in Figs. 10.7 and 10.8 in the XY plane is $(0.2046144211, 0)$. Note that there are more cusps in Fig. 10.7 than in Fig. 10.5 but these are smaller, and that the trajectory is becoming more circular.

Example 10.3
Set $V = 16$, $\alpha = 15°$. Figure 10.9 shows a relatively circular trajectory for P_1 in the XY plane using all 2200 points, which yield just over 1 cycle. The point $\bar{P}$ for the entire trajectory lies on the line $(0.2046144211, 0, z)$ with $\bar{z}$ in the range $0.763 < \bar{z} < 0.764$.

Example 10.4
Set $V = 4$, $\alpha = 30°$. Figure 10.10 shows the cusped path of P_1 in the XY plane using the first 350 points of its trajectory, which yields just over 1 complete cycle. There are 6^+ cycles in the 2200 point trajectory shown in Fig. 10.11, which has been enlarged for clarity. The point $\bar{P}$ for the entire 2200 point trajectory lies on the line $(0.3952847076, 0, z)$, with $\bar{z}$ rising and falling in the range $0.652 < \bar{z} < 0.685$.

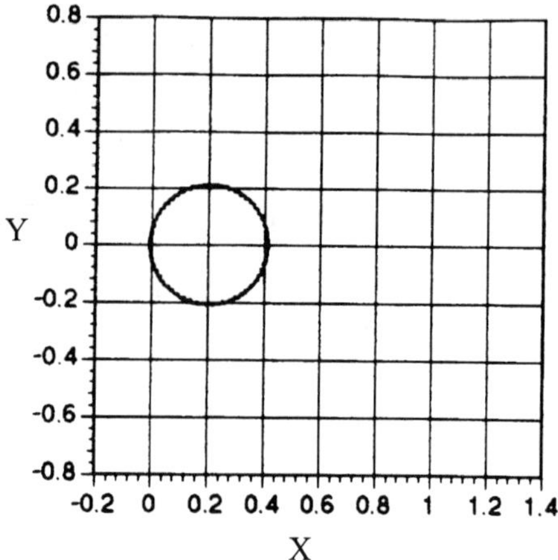

Figure 10.7.

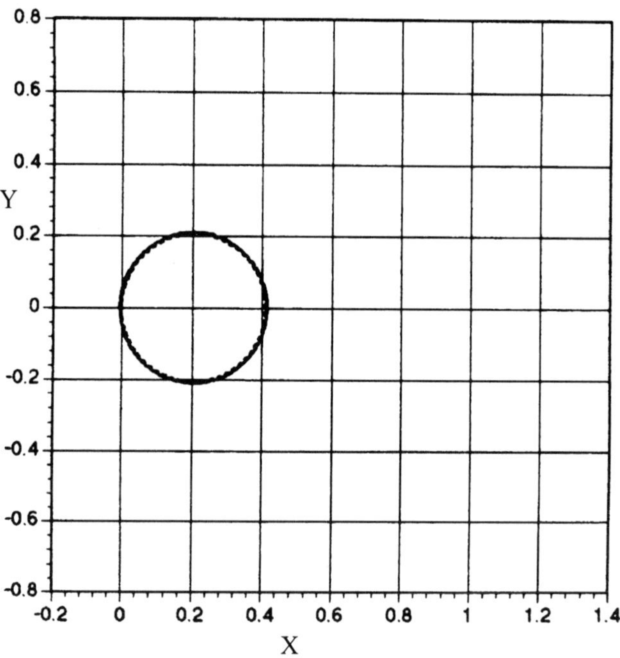

Figure 10.8.

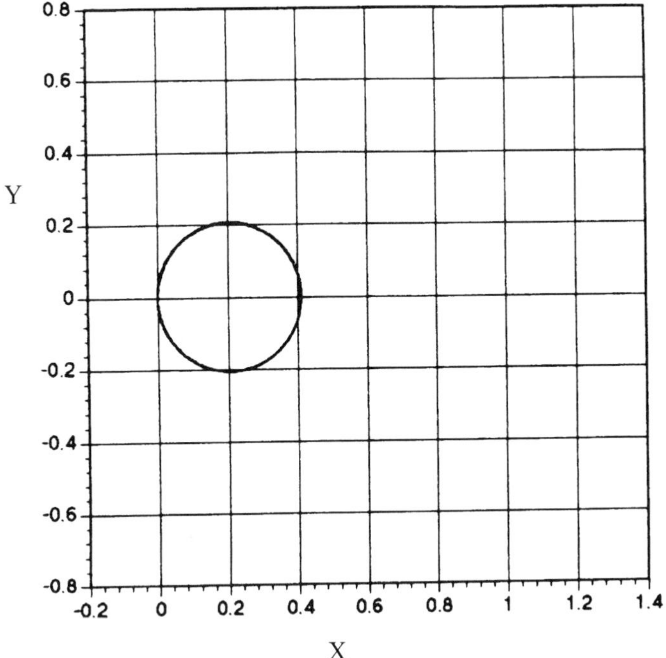

Figure 10.9.

Example 10.5

Set $V = 8$, $\alpha = 30°$. Figure 10.12 shows the cusped path of P_1 in the XY plane for the first 750 points of its trajectory, which yields just over 1 complete cycle. There are 3^+ cycles in the entire 2200 point trajectory shown in Fig. 10.13. The point $\bar{P}$ for the entire 2200 point trajectory lies on the line $(0.3952847076, 0, z)$, with $\bar{z}$ rising and falling in the range $0.678 < \bar{z} < 0.685$. Compared to Example 4, the number of cusps is increasing but their size is decreasing.

Example 10.6

Set $V = 16$, $\alpha = 30°$. Figure 10.14 shows a relatively circular path for P_1 using all 2200 trajectory points, which yield just over 1 cycle. The center of the circle is $(0.3952847076, 0, z)$. The point $\bar{P}$ is always on the line $(0.3952847076, 0, Z)$, with $\bar{z}$ in the range $0.683 < \bar{z} < 0.685$.

Example 10.7

Set $V = 4$, $\alpha = 45°$. Figure 10.15 shows the cusped path of P_1 in the XY plane using the first 350 points of its trajectory, which yields just over 1 complete cycle. There are 6^+ cycles in the 2200 point trajectory shown in Figure 10.16. The point

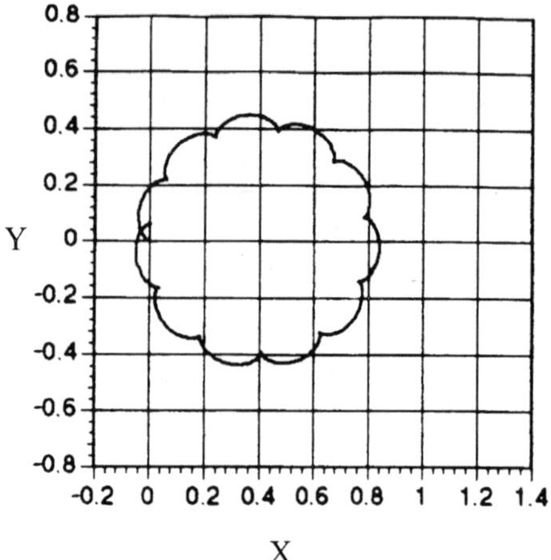

Figure 10.10.

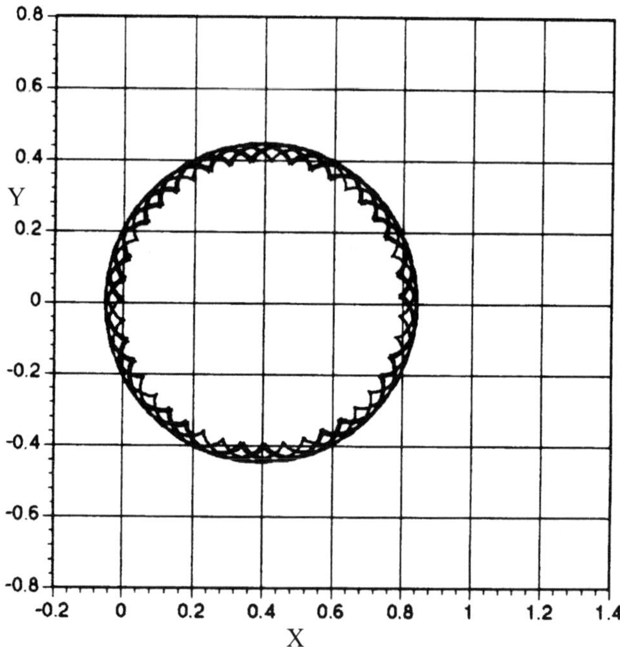

Figure 10.11.

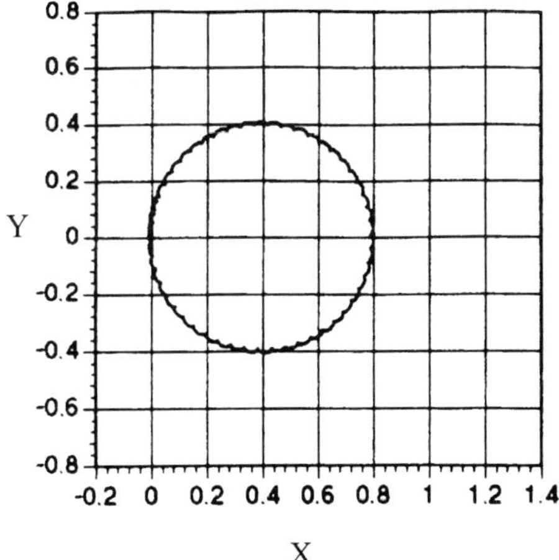

Figure 10.12.

$\bar{P}$ always lies on the line $(0.5590169945, 0, z)$, with $\bar{z}$ rising and falling in the range $0.497 < \bar{z} < 0.559$.

Example 10.8
Set $V = 8$, $\alpha = 45°$. Figure 10.17 shows the cusped path of P_1 in the XY plane for the first 750 point of its trajectory, which yields just over 1 complete cycle. There are 3^+ cycles in the entire 2200 point trajectory shown in Fig. 10.18. The point $\bar{P}$ is always on the line $(0.5590169945, 0, z)$, with $\bar{z}$ rising and falling in the range $0.545 < \bar{z} < 0.559$.

Example 10.9
Set $V = 16$, $\alpha = 45°$. Figure 10.19 shows the relatively circular trajectory of P_1 in the XY plane for all 2200 points, which yields just over 1 cycle. The point $\bar{P}$ is always on the line $(0.5590169945, 0, z)$, with $\bar{z}$ in the range of $0.555 < \bar{z} < 0.559$.

Example 10.10
$V = 1$, $\alpha = 15°$. The top falls to the XY plane.

Examples 1–9 reveal that each trajectory has a center point, that the trajectories become more circular with increasing V, that the variation in $\bar{z}$ diminishes as V increases, and that the diameter of the trajectory increases with α.

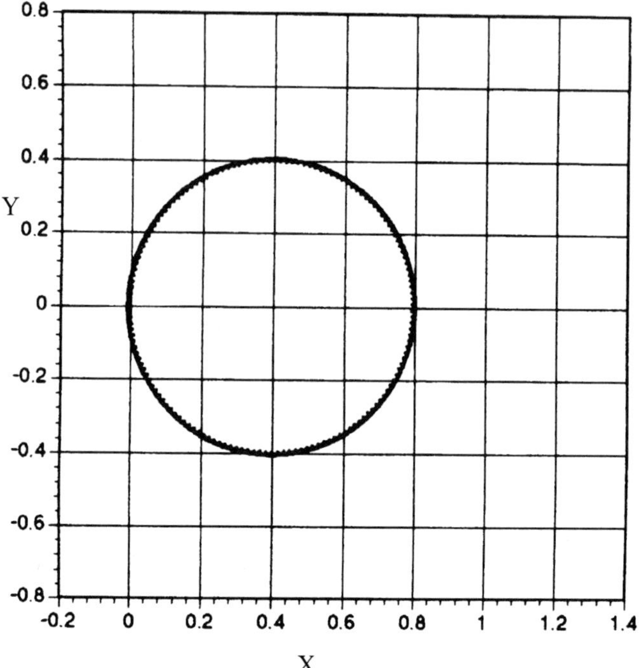

Figure 10.13.

We turn next to the difficult problem (Gray (1959)) of a *nonhomogenous* top.
Note that in earlier sections we used the term *geometric center* throughout rather
than *mass center* because of the examples to be considered next.

Example 10.11
Let $V = 4$, $\alpha = 15°$, as in Example 1, but set $m_2 = 0.995$. The resulting 2200
point trajectory for P_1 in the XY plane is shown in Fig. 10.20. It shows cusped
motion which is similar to that in Fig. 10.5 but is in motion to the left. The motion
of $\bar{P}$ is fully three dimensional with the projection of its first 100 points shown in
Fig. 10.21. Though $\bar{y}$ and $\bar{z}$ show only small variations, it is $\bar{x}$ that shows significant
motion.

Example 10.12
Let $V = 4$, $\alpha = 30°$, as in Example 4, but set $m_2 = 1.005$. The results are similar
to those in Example 11 and are shown in Figs. 10.22 and 10.23. However, this
time, the motion of P_1 in the XY plane is to the right.

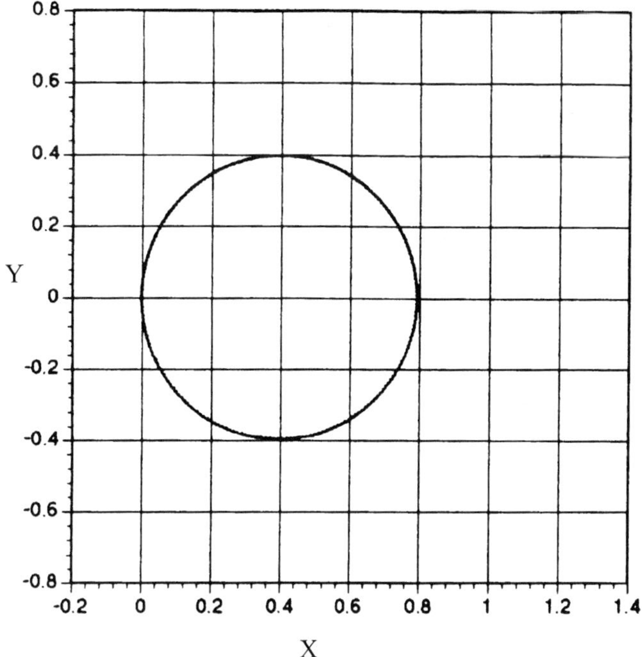

Figure 10.14.

Example 10.13

Let $V = 8$, $\alpha = 30°$, as in Example 5, but set $m_2 = 1.005$ and $m_3 = 0.95$. The entire motion for P_1 is shown in Fig. 10.24 and the projection for $\bar{P}$, but for only 100 points, is shown in Fig. 10.25. Figure 10.24 shows complex looping motion up and to the right.

Example 10.14

Let $V = 4$, $\alpha = 45°$, $m_2 = 0.995$, $m_3 = 0.985$. The resulting motion of P_1 for the entire 2200 points is shown in Fig. 10.26. The projected motion of $\bar{P}$ for the first 100 points is shown in Fig. 10.27. The motion of P_1 is up and to the right, but similar to that of Fig. 10.22.

With regard to Examples 11–14 and the corresponding graphs in Figs. 10.21, 10.23, 10.25, and 10.27, it is worth noting that these graphs, with only 100 points, characterize completely the entire graphs with 2200 points.

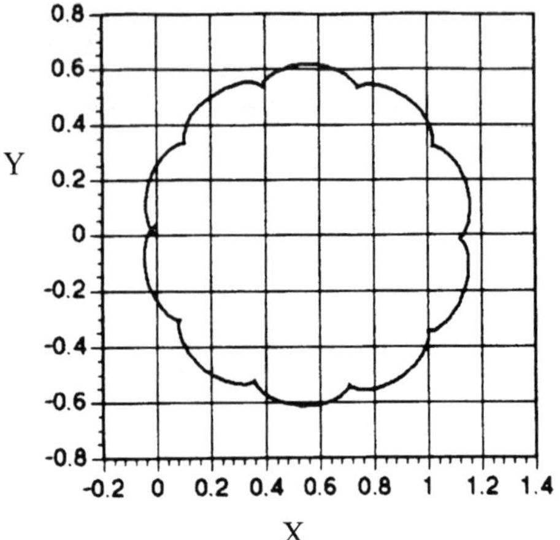

Figure 10.15.

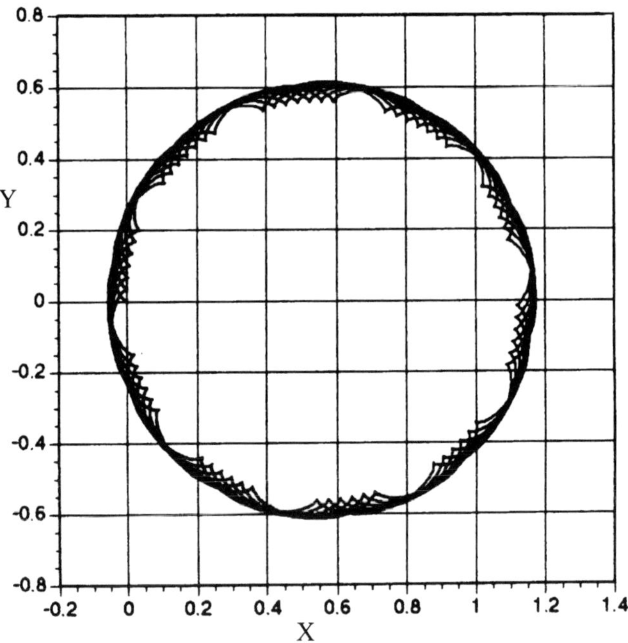

Figure 10.16.

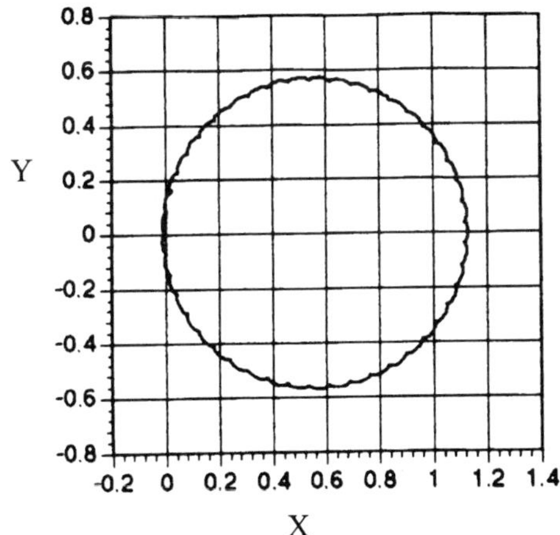

Figure 10.17.

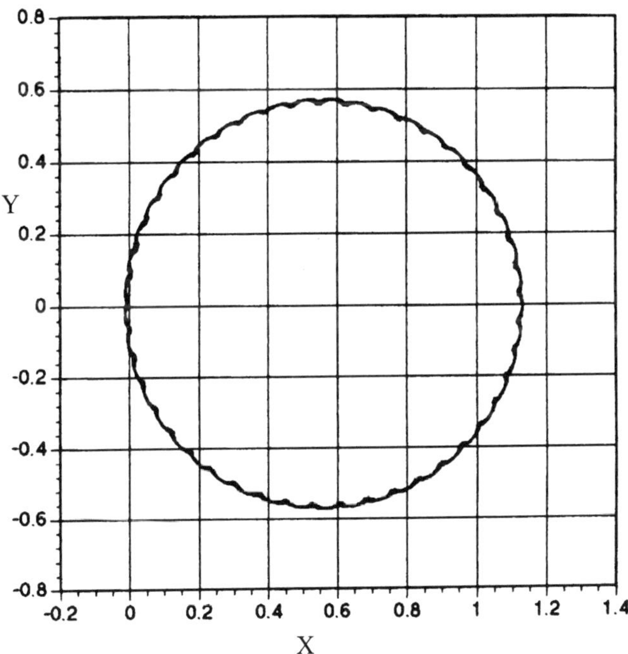

Figure 10.18.

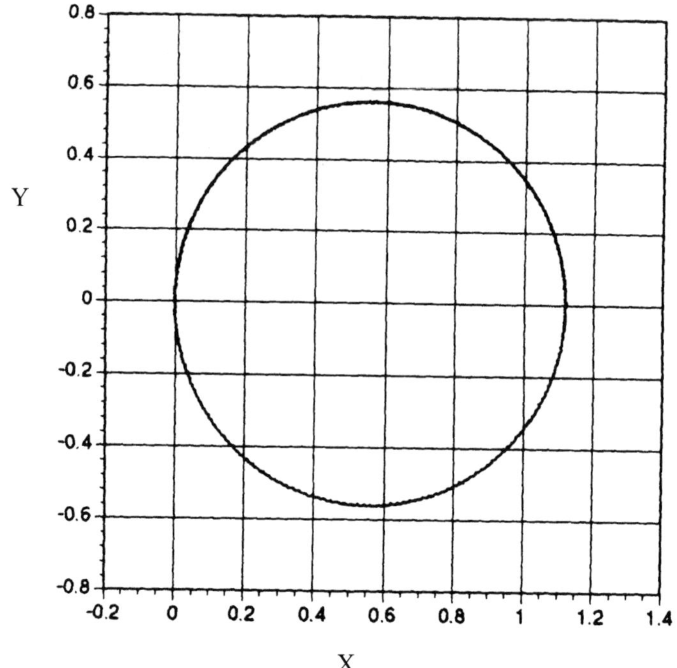

Figure 10.19.

10.6 Extensions

In order to extend the present discussion to more complex rigid bodies, one can proceed as follows. A given body should be decomposed first into regular tetrahedral building blocks. Let the resulting vertices be P_1, P_2, $\ldots$, P_N. Molecular type force formulas between pairs of particles can then be applied, but in a fashion so that each particle interacts only with nearest neighbors. This will preserve rigidity. If a particular material composition is specified, then the total mass need only be distributed equally over P_1, P_2, $\ldots$, P_N. Finally, the resulting N-body problem can be solved in a fashion entirely analogous to that described in Sect. 10.4.

Note also that any other conservative formula, different from (10.16) which ensures rigidity, may also be used.

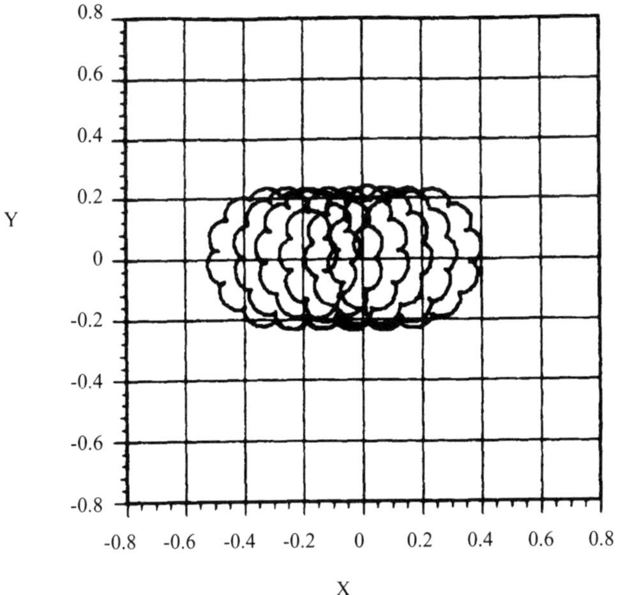

Figure 10.20.

10.7 A Discrete, Rigid Hexahedral Gyroscope

We turn next to the simulation of gyroscopic motion. As shown in Fig. 10.28, consider a hexahedron with six congruent, regular triangular faces. Let the edge length of each triangle be R and let the vertices be $P_i(x_i, y_i, z_i)$, $i = 1, 2, 3, 4, 5$. Initially, let

$$(x_1, y_1, z_1) = (0, 0, 0), \quad (x_2, y_2, z_2) = (0, R\sqrt{3}/3, R\sqrt{6}/3)$$

$$(x_3, y_3, z_3) = \left(\frac{1}{2}R, -R\sqrt{3}/6, R\sqrt{6}/3\right)$$

$$(x_4, y_4, z_4) = \left(-\frac{1}{2}R, -R\sqrt{3}/6, R\sqrt{6}/3\right), \quad (x_5, y_5, z_5) = (0, 0, 2R\sqrt{6}/3).$$

The geometric center $\bar{P} = (\bar{x}, \bar{y}, \bar{z})$ of the hexahedron is $(0, 0, R\sqrt{6}/3)$.

We define the neighbors of P_1 and P_5 to be P_2, P_3, P_4. The neighbors of any one of P_2, P_3, P_4 are defined to be *all* the other vertices. Thus, P_1 and P_5 have three neighbors each, while P_2, P_3 and P_4 have four.

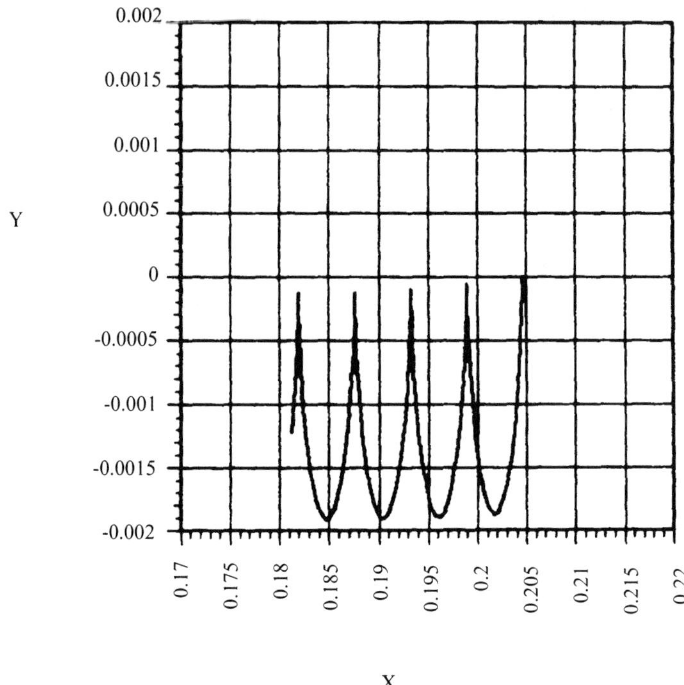

Figure 10.21.

Next, let us set P_2, P_3, P_4 in rotation in the plane $z = R\sqrt{6}/3$. As in Fig. 10.3, let the velocity of each particle be perpendicular to the line joining that particle to the geometric center $\bar{P}$. Assume that P_2, P_3, P_4 each has the same speed V, so that we take the initial velocities $\vec{v}_i = (v_{ix}, v_{iy}, v_{iz})$, $i = 1, 2, 3, 4, 5$, of each P_i to be

$$\vec{v}_1 = (0, 0, 0), \quad \vec{v}_2 = (V, 0, 0), \quad \vec{v}_3 = \left(-\frac{1}{2}V, -V\sqrt{3}/2, 0 \right)$$

$$\vec{v}_4 = \left(-\frac{1}{2}V, V\sqrt{3}/2, 0 \right), \quad \vec{v}_5 = (0, 0, 0).$$

Finally, we want the rotating hexahedron to be tilted, initially, relative to the Z axis, so that we assume that the line joining P_1 to the geometric center $\bar{P}$ forms an angle α relative to the Z axis. As in Fig. 10.4, this will be done by rotating the XZ plane through an angle α. Thus, the new positions (x_i', y_i', z_i') and the new velocities $(v_{ix'}, v_{iy'}, v_{iz'})$, $i = 1, 2, 3, 4, 5$, satisfy

$$x_i' = x_i \cos\alpha + z_i \sin\alpha, \quad y_i' = y_i, \quad z_i' = -x_i \sin\alpha + z_i \cos\alpha,$$

$$v_{ix'} = v_{ix} \cos\alpha + v_{iz} \sin\alpha, \quad v_{iy'} = v_{iy}, \quad v_{iz'} = -v_{ix} \sin\alpha + v_{iz} \cos\alpha.$$

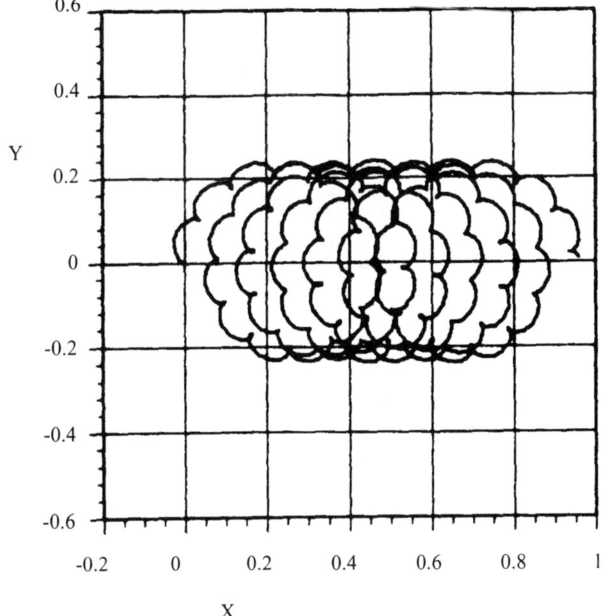

Figure 10.22.

Thus, once the parameters R, V and α are given, all initial data for a tilted, rotating, hexahedral gyroscope are determined.

Note, however, that a fundamental difference in our present approach from that of Sect. 10.3, is that we now assume that P_1 is fixed for all time, as is the case for gyroscopic motion.

10.8 Dynamical Equations

The motion of our rotating gyroscope is now treated as a five-body problem. At any time t, let P_i, $i = 1, 2, 3, 4, 5$, be located at $\vec{r}_i = (x_i, y_i, z_i)$, have velocity $\vec{v}_i = (\dot{x}_i, \dot{y}_i, \dot{z}_i) = (v_{ix}, v_{iy}, v_{iz})$, and have acceleration $\vec{a}_i = (\ddot{x}_i, \ddot{y}_i, \ddot{z}_i) = (\dot{v}_{ix}, \dot{v}_{iy}, \dot{v}_{iz})$. Let the mass of each P_i be m_i. For $i \neq j$, let $\vec{r}_{ij}$ be the vector from P_i to P_j and let r_{ij} be the magnitude of $\vec{r}_{ij}$, $i = 1, 2, 3, 4, 5$; $j = 1, 2, 3, 4, 5$; $i \neq j$. Let $\phi = \phi(r_{ij})$ be a potential function defined by the pair P_i, P_j, $i \neq j$. Then for $i = 1, 2, 3, 4, 5$, the Newtonian dynamical equations for the motion of

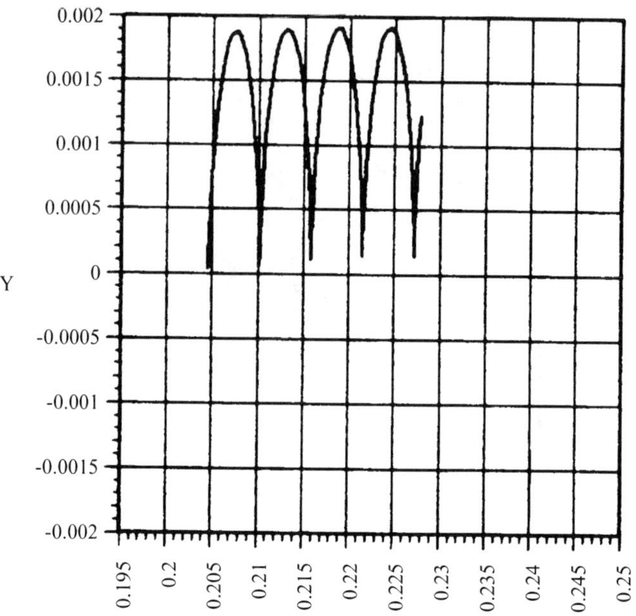

Figure 10.23.

the particles are the second-order differential equations

$$m_i a_{ix} = -\frac{\partial\phi}{\partial r_{ij}}\frac{x_i - x_j}{r_{ij}} - \frac{\partial\phi}{\partial r_{ik}}\frac{x_i - x_k}{r_{ik}}$$

$$-\frac{\partial\phi}{\partial r_{im}}\frac{x_i - x_m}{r_{im}} - \delta_i\frac{\partial\phi}{\partial r_{in}}\frac{x_i - x_n}{r_{in}} \qquad (10.18)$$

$$m_i a_{iy} = -\frac{\partial\phi}{\partial r_{ij}}\frac{y_i - y_j}{r_{ij}} - \frac{\partial\phi}{\partial r_{ik}}\frac{y_i - y_k}{r_{ik}}$$

$$-\frac{\partial\phi}{\partial r_{im}}\frac{y_i - y_m}{r_{im}} - \delta_i\frac{\partial\phi}{\partial r_{in}}\frac{y_i - y_n}{r_{in}} \qquad (10.19)$$

$$m_i a_{iz} = -\frac{\partial\phi}{\partial r_{ij}}\frac{z_i - z_j}{r_{ij}} - \frac{\partial\phi}{\partial r_{ik}}\frac{z_i - z_k}{r_{ik}}$$

$$-\frac{\partial\phi}{\partial r_{im}}\frac{z_i - z_m}{r_{im}} - \delta_i\frac{\partial\phi}{\partial r_{in}}\frac{z_i - z_n}{r_{in}} - g_i. \qquad (10.20)$$

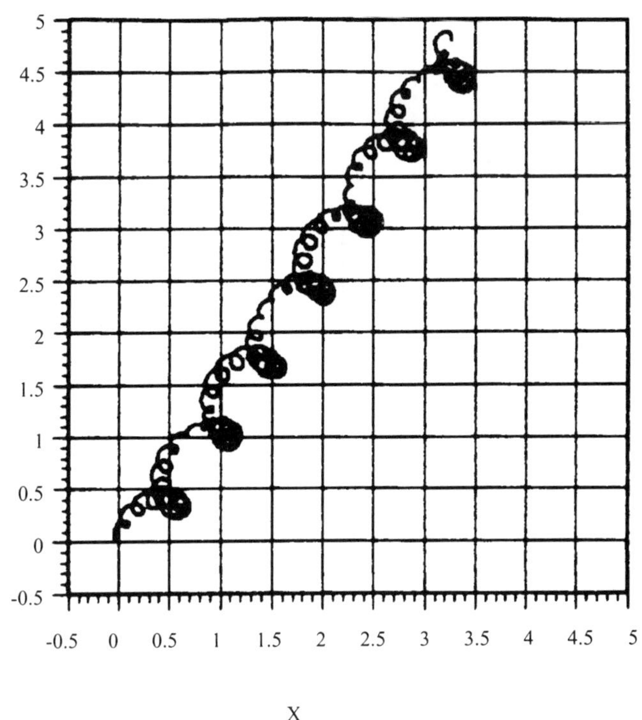

Figure 10.24.

In (10.18)–(10.20), $\delta_1 = \delta_5 = 0$, $\delta_2 = \delta_3 = \delta_4 = 1$. Moreover, if $i = 1$, then $j = 2, k = 3, m = 4$; if $i = 2$ then $j = 1, k = 3, m = 4, n = 5$; if $i = 3$, then $j = 1, k = 2, m = 4, n = 5$; if $i = 4$, then $j = 1, k = 2, m = 3, n = 5$; if $i = 5$, then $j = 2, k = 3, m = 4$. Moreover, we choose $g_1 = 0$, $g_2 = g_3 = g_4 = g_5 > 0$, so that gravity acts only on P_2, P_3, P_4, P_5.

Equations (10.18)–(10.20) are fully conservative.

10.9 Numerical Method

In order to solve system (10.18)–(10.20) in a fashion which conserves exactly the same energy, linear momentum and angular momentum, we first rewrite it as the

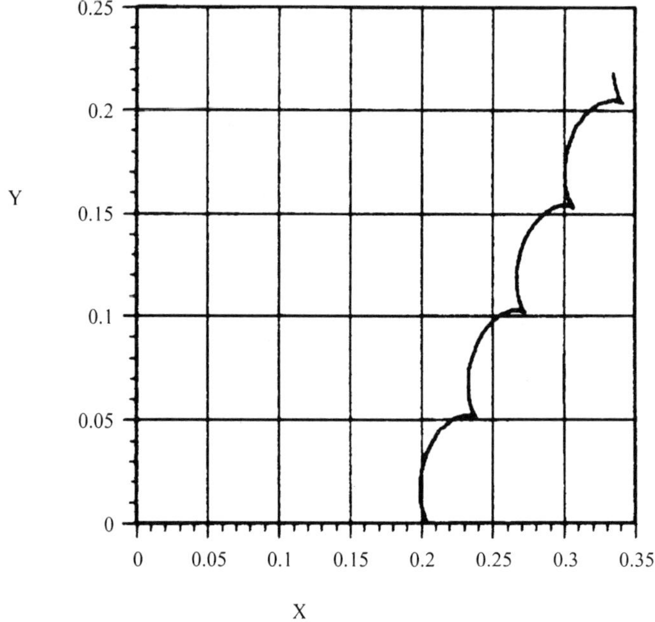

Figure 10.25.

following equivalent first order system:

$$\frac{dx_i}{dt} = v_{ix} \tag{10.21}$$

$$\frac{dy_i}{dt} = v_{iy} \tag{10.22}$$

$$\frac{dz_i}{dt} = v_{iz} \tag{10.23}$$

$$m_i \frac{dv_{ix}}{dt} = -\frac{\partial \phi}{\partial r_{ij}} \frac{x_i - x_j}{r_{ij}} - \frac{\partial \phi}{\partial r_{ik}} \frac{x_i - x_k}{r_{ik}}$$
$$- \frac{\partial \phi}{\partial r_{im}} \frac{x_i - x_m}{r_{im}} - \delta_i \frac{\partial \phi}{\partial r_{in}} \frac{x_i - x_n}{r_{in}} \tag{10.24}$$

$$m_i \frac{dv_{iy}}{dt} = -\frac{\partial \phi}{\partial r_{ij}} \frac{y_i - y_j}{r_{ij}} - \frac{\partial \phi}{\partial r_{ik}} \frac{y_i - y_k}{r_{ik}}$$
$$- \frac{\partial \phi}{\partial r_{im}} \frac{y_i - y_m}{r_{im}} - \delta_i \frac{\partial \phi}{\partial r_{in}} \frac{y_i - y_n}{r_{in}} \tag{10.25}$$

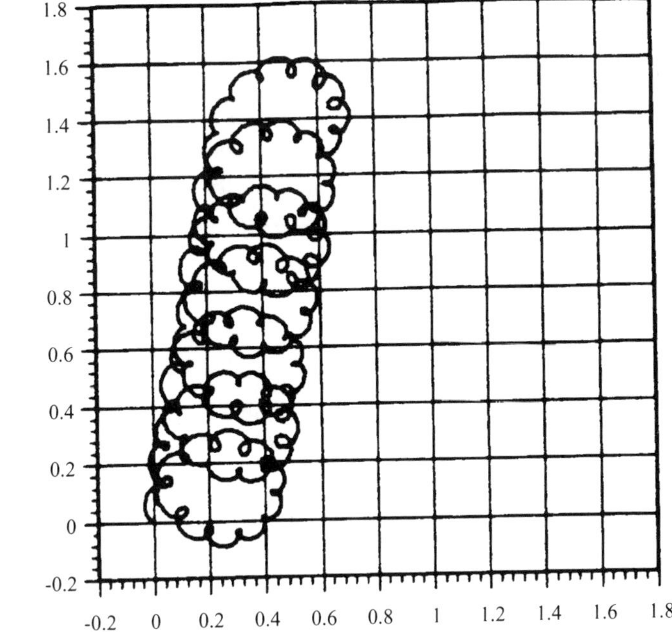

Figure 10.26.

$$m_i \frac{dv_{iz}}{dt} = -\frac{\partial \phi}{\partial r_{ij}} \frac{z_i - z_j}{r_{ij}} - \frac{\partial \phi}{\partial r_{ik}} \frac{z_i - z_k}{r_{ik}}$$

$$-\frac{\partial \phi}{\partial r_{im}} \frac{z_i - z_m}{r_{im}} - \delta_i \frac{\partial \phi}{\partial r_{in}} \frac{z_i - z_n}{r_{in}} - g_i. \tag{10.26}$$

We now choose difference equation approximations for (10.21)–(10.26). For a fixed time step Δt, let $t_N = N\Delta t$, $N = 0, 1, 2, 3, \ldots$. At t_N let P_i be at $\vec{r}_{i.N} = (x_{i.N}, y_{i.N}, z_{i.N})$ with velocity $\vec{v}_{i.N} = (v_{i.x.N}, v_{i.y.N}, v_{i.z.N})$. Then (10.21)–(10.26) will be approximated by

$$\frac{x_{i.N+1} - x_{i.N}}{\Delta t} = \frac{v_{i.x.N+1} + v_{i.x.N}}{2} \tag{10.27}$$

$$\frac{y_{i.N+1} - y_{i.N}}{\Delta t} = \frac{v_{i.y.N+1} + v_{i.y.N}}{2} \tag{10.28}$$

$$\frac{z_{i.N+1} - z_{i.N}}{\Delta t} = \frac{v_{i.z.N+1} + v_{i.z.N}}{2} \tag{10.29}$$

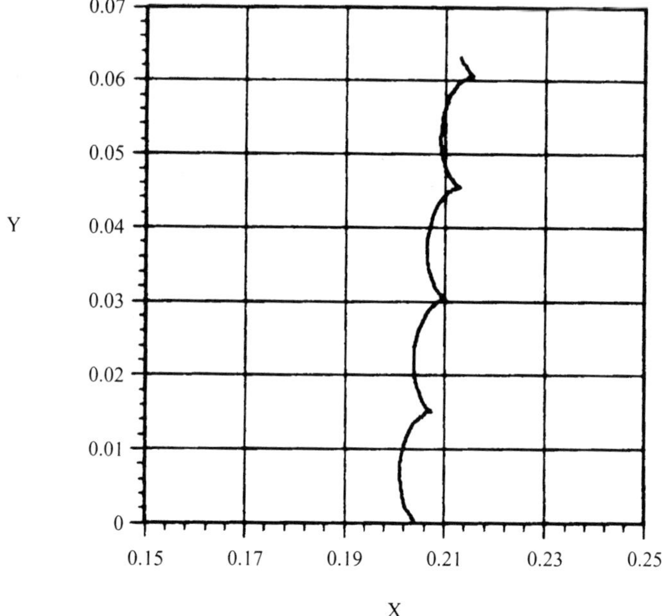

Figure 10.27.

$$m_i \frac{v_{i.x.N+1} - v_{i.x.N}}{\Delta t} = -\frac{\phi(r_{ij.N+1} - \phi(r_{ij.N})}{r_{ij.N+1} - r_{ij.N}} \cdot \frac{x_{i.N+1} + x_{i.N} - x_{j.N+1} - x_{j.N}}{r_{ij.N+1} + r_{ij.N}}$$

$$- \frac{\phi(r_{ik.N+1}) - \phi(r_{ik.N})}{r_{ik.N+1} - r_{ik.N}} \cdot \frac{x_{i.N+1} + x_{i.N} - x_{k.N+1}) - x_{k.N}}{r_{ik.N+1} + r_{ik.N}}$$

$$- \frac{\phi(r_{im.N+1}) - \phi(r_{im.N})}{r_{im.N+1} - r_{im.N}} \cdot \frac{x_{i.N+1} + x_{i.N} - x_{m.N+1} - x_{m.N}}{r_{im.N+1} + r_{im.N}}$$

$$- \delta_i \frac{\phi(r_{in.N+1}) - \phi(r_{in.N})}{r_{in.N+1} - r_{in.N}} \cdot \frac{x_{i.N+1} + x_{i.N} - x_{n.N+1} - x_{n.N}}{r_{in.N+1} + r_{in.N}} \qquad (10.30)$$

$$m_i \frac{v_{i.y.N+1} - v_{i.y.N}}{\Delta t} = -\frac{\phi(r_{ij.N+1}) - \phi(r_{ij.N})}{r_{ij.N+1} - r_{ij.N}} \cdot \frac{y_{i.N+1} + y_{i.N} - y_{j.N+1} - y_{j.N}}{r_{ij.N+1} + r_{ij.N}}$$

$$- \frac{\phi(r_{ik.N+1}) - \phi(r_{ik.N})}{r_{ik.N+1} - r_{ik.N}} \cdot \frac{y_{i.N+1} + y_{i.N} - y_{k.N+1} - y_{k.N}}{r_{ik.N+1} + r_{ik.N}}$$

$$- \frac{\phi(r_{im.N+1}) - \phi(r_{im.N})}{r_{im.N+1} - r_{im.N}} \cdot \frac{y_{i.N+1} + y_{i.N} - y_{m.N+1} - y_{m.N}}{r_{im.N+1} + r_{im.N}}$$

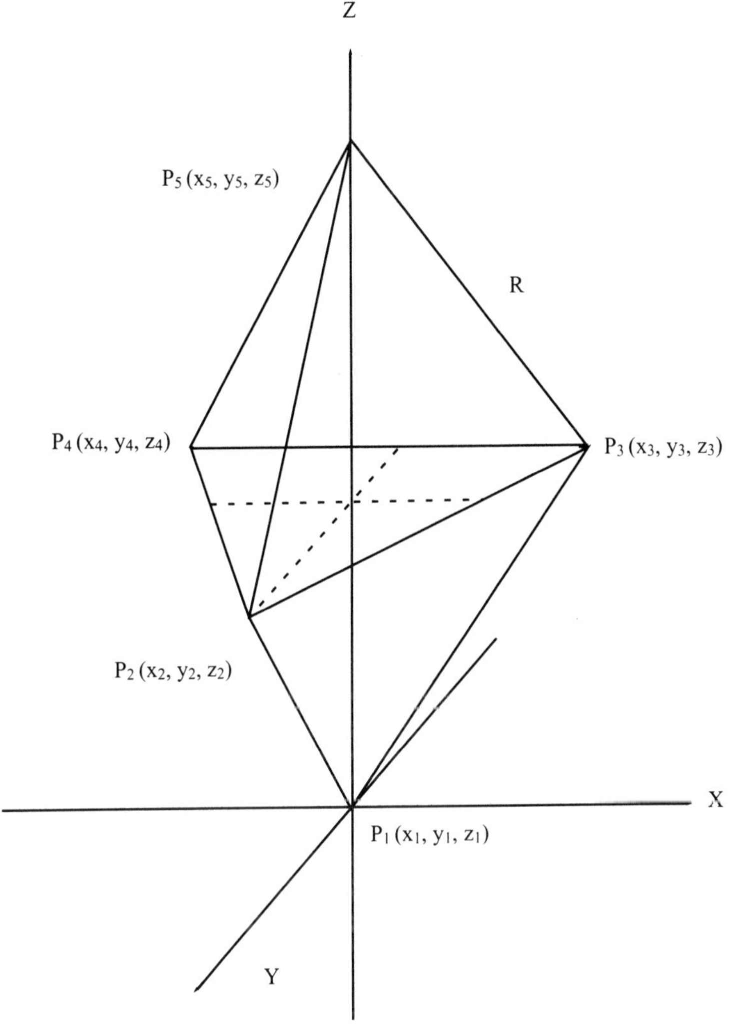

Figure 10.28.

$$- \delta_i \frac{\phi(r_{in,N+1}) - \phi(r_{in,N})}{r_{in,N+1} - r_{in,N}} \cdot \frac{y_{i,N+1} + y_{i,N} - y_{n,N+1} - y_{n,N}}{r_{in,N+1} + r_{in,N}} \qquad (10.31)$$

$$m_i \frac{v_{i,z,N+1} - v_{i,z,N}}{\Delta t} = - \frac{\phi(r_{ij,N+1} - \phi(r_{ij,N})}{r_{ij,N+1} - r_{ij,N}} \cdot \frac{z_{i,N+1} + z_{i,N} - z_{j,N+1} - z_{j,N}}{r_{ij,N+1} + r_{ij,N}}$$

$$- \frac{\phi(r_{ik,N+1}) - \phi(r_{ik,N})}{r_{ik,N+1} - r_{ik,N}} \cdot \frac{z_{i,N+1} + z_{i,N} - z_{k,N+1} - z_{k,N}}{r_{ik,N+1} + r_{ik,N}}$$

$$- \frac{\phi(r_{im,N+1}) - \phi(r_{im,N})}{r_{im,N+1} - r_{im,N}} \cdot \frac{z_{i,N+1} + z_{i,N} - z_{m,N+1} - z_{m,N}}{r_{im,N+1} + r_{im,N}}$$

$$- \delta_i \frac{\phi(r_{in,N+1}) - \phi(r_{in,N})}{r_{in,N+1} - r_{in,N}} \cdot \frac{z_{i,N+1} + z_{i,N} - z_{n,N+1} - z_{n,N}}{r_{in,N+1} + r_{in,N}} - g_i. \qquad (10.32)$$

Difference equations (10.27)–(10.32) are consistent with the differential system (10.21)–(10.26) and conserve exactly the same system invariants. For each of $N = 0, 1, 2, 3, \ldots$, system (10.27)–(10.32) consists of 30 equations in the unknowns $x_{i,N+1}, y_{i,N+1}, z_{i,N+1}, v_{i,x,N+1}, v_{i,y,N+1}, v_{i,z,N+1}, i = 1, 2, 3, 4, 5$.

In all the examples to be described, the results will be given relative to P_1, whose position and velocity coordinates at the end of each time step are set to zero. As a consequence, the motion of the hexahedron's geometric center $\bar{P}$ fully characterizes the motion of the gyroscope.

Typical FORTRAN programs for generating initial data and for trajectory calculation are given in Appendix A4.

10.10 Examples

In considering examples, we must first choose a potential function, and, as in Sect. 10.5, we choose again

$$\phi = A \left[-\frac{1}{r_{ij}^3} + \frac{1}{r_{ij}^5} \right], \quad A > 0. \qquad (10.33)$$

From (10.33), it follows again that

$$F = A \left[-\frac{3}{r_{ij}^4} + \frac{5}{r_{ij}^6} \right].$$

Thus, $F(\bar{r}) = 0$ provided $\bar{r} = (5/3)^{\frac{1}{2}} \sim 1.290994449$.

We now choose the parameter R to be

$$R = 1.290994449. \qquad (10.34)$$

For this value of R the force between any two neighboring particles is zero, so that the gyroscope is physically stable. The parameters A, g and m_i are scaled for computational convenience to be $A = 10^6$, $g = 0.980$, $m_i = 1$, $i = 1, 2, 3, 4, 5$, unless otherwise indicated. The time step Δt is chosen to be $\Delta t = 10^{-5}$. In the figures to be given, units on the Z axis are often rescaled to accentuate the character of the resulting trajectory. In all the calculations, the distance between any two neighbors is always 1.291.

In this first example, let $V = 20$, $\alpha = 8°$. Figure 10.29 shows the resulting circular precession of the geometric center every 10,000 time steps through

28,000,000 time steps. The graph consists of 2800 points. $\bar{P}$ has completed just over one complete rotation. The $\bar{z}$ values vary in the narrow range $1.0435 < \bar{z} < 1.04395$. Figure 10.30 shows the projection in the XY plane of the motion shown in Fig. 10.29.

In the next six examples, $V = 15$, but α varies successively through the values $15°$, $30°$, $45°$, $60°$, $75°$, $90°$. The results are shown in Figs. 10.31–10.36. The figures reveal precession of the geometric center with the radius of the trajectory increasing with α. For $\alpha = 90°$, one finds $-0.006 < \bar{z} < 0.000$.

For the next example, set $V = 3$, $\alpha = 16°$. The resulting motion of the geometric center is the nutation shown in Fig. 10.37. In this example $\bar{z}$ varies in the range $0.4 < \bar{z} < 0.54$, so that the gyroscope displays a periodic rising and falling motion. The projection of the trajectory in the XY plane is shown in Fig. 10.38. The figures indicate clearly the formation of cusps.

For our final example, choose again $V = 15$ and $\alpha = 15°$, but let the mass m_2 of P_2 be 3. Thus, the resulting hexahedron, though symmetric, is nonhomogenous. The resulting trajectory of $\bar{P}$ is shown every 10,000 time steps through 30,000,000 time steps in Fig. 10.39. $\bar{P}$ has completed just over one complete rotation in this time and the graph contains 3000 points. That this graph is different from all previous graphs is quite visible. To analyze the resulting behavior in greater detail, we first examined the projection of the trajectory in the XY plane, since the

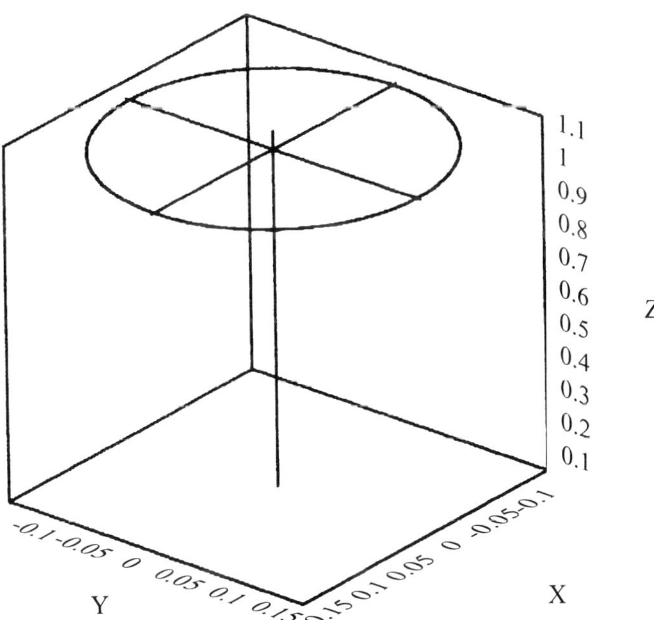

Figure 10.29.

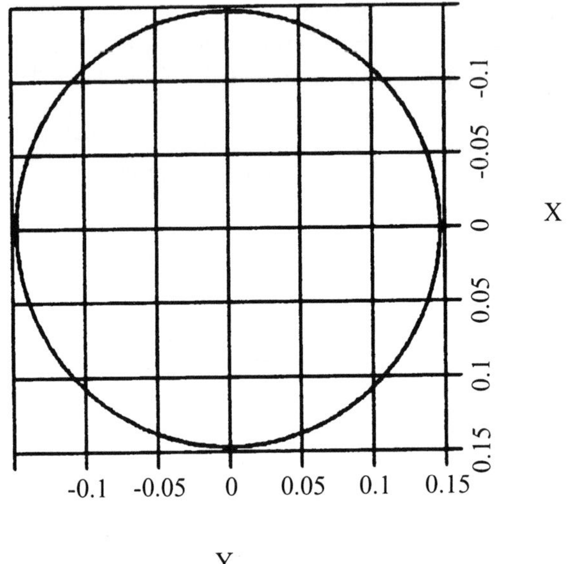

Figure 10.30.

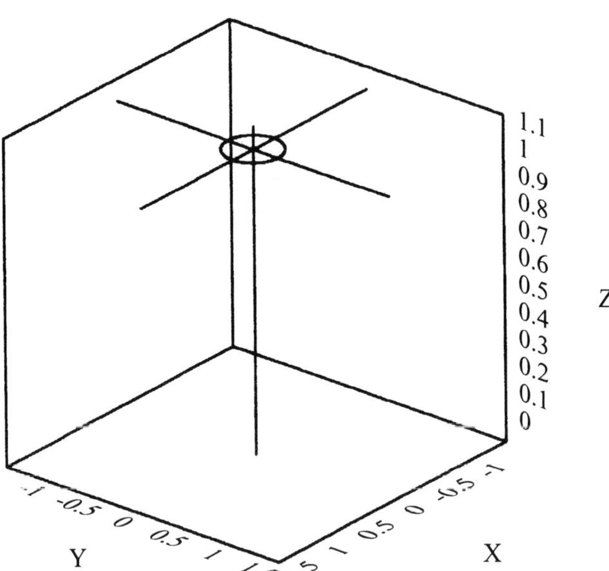

Figure 10.31.

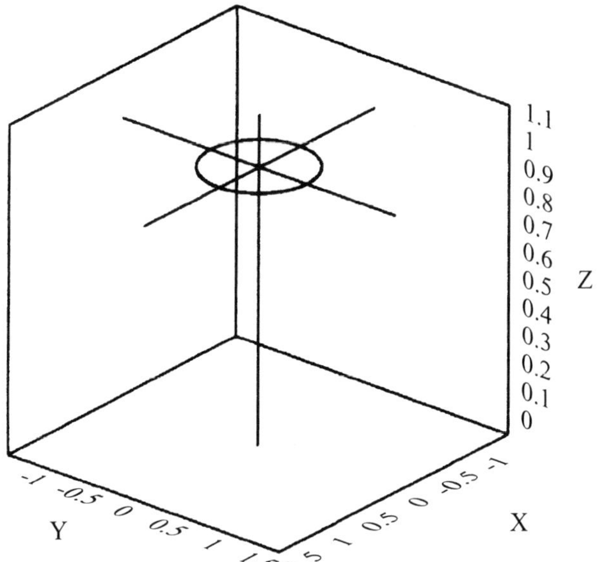

Figure 10.32.

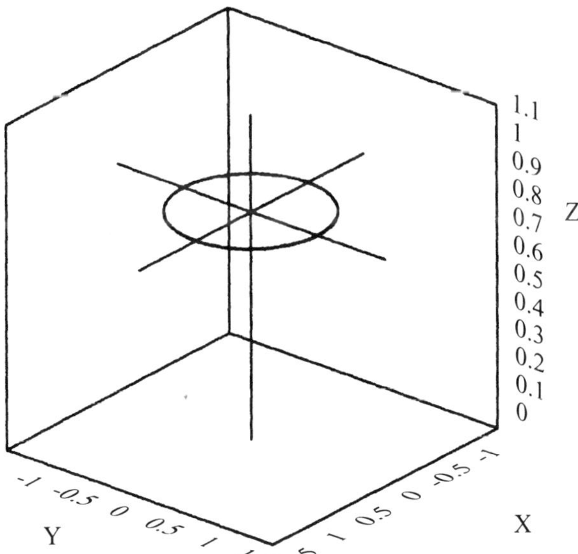

Figure 10.33.

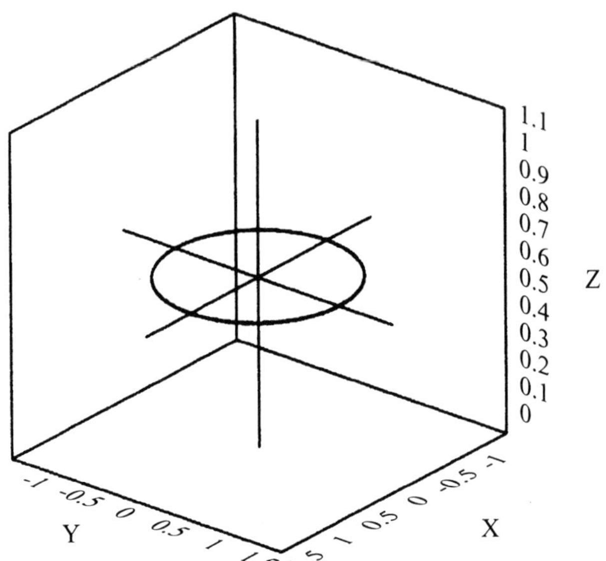

Figure 10.34.

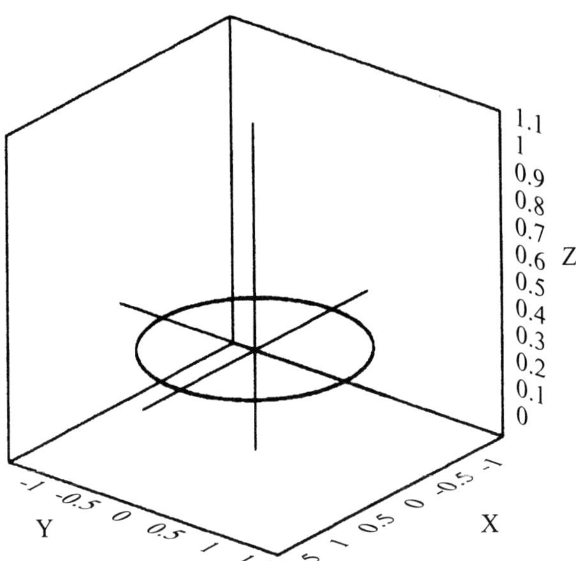

Figure 10.35.

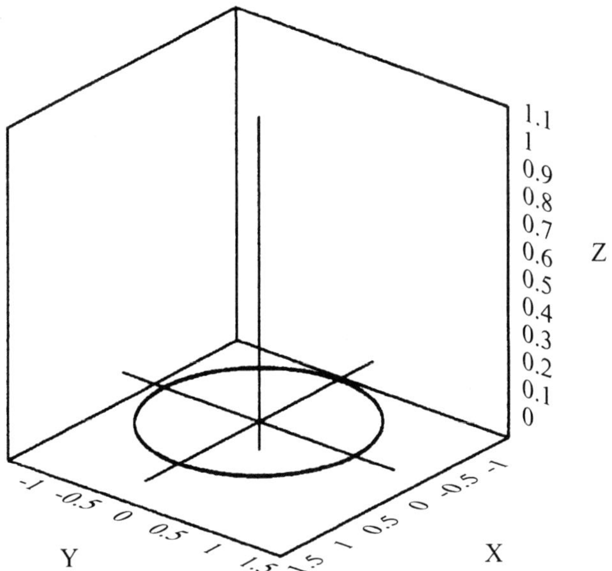

Figure 10.36.

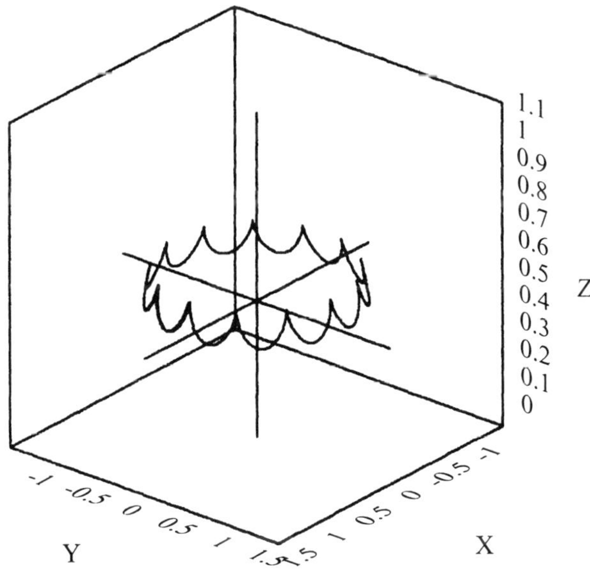

Figure 10.37.

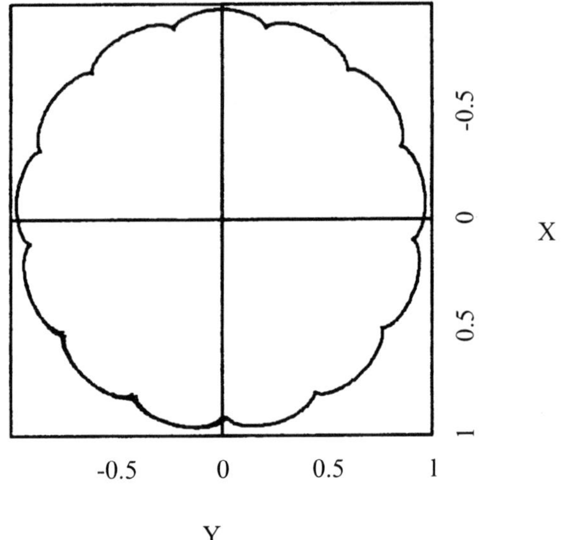

Figure 10.38.

variation in $\bar{z}$ was entirely in the small range $1.0155 < \bar{z} < 1.0205$. This projection is shown in Fig. 10.40 and reveals that the variations of X and Y must be relatively complex. To study these variations more precisely, the motion was simulated over its first 500,000 steps and plotted every 500 time steps. The projection of the resulting 1000 points in the XY plane is shown in Fig. 10.41. What appears to be happening is that the X and Y coordinates are looping. However, plotting every 500 points has resulted in a graph with polygonal shapes rather than smooth shapes. To analyze the motion further, we concentrated only on the lower right corner of the graph shown in Fig. 10.41. For this purpose we reran the trajectory but for only 150,000 time steps and plotted every 50 time steps. The projected motion in the XY plane is shown in Fig. 10.42 and contains 3000 points. What is revealed is both looping and cusp formation, which was totally unexpected.

10.11 Remark

Small perturbations of the parameters used for the results shown in Figs. 10.29–10.38 resulted in entirely similar results.

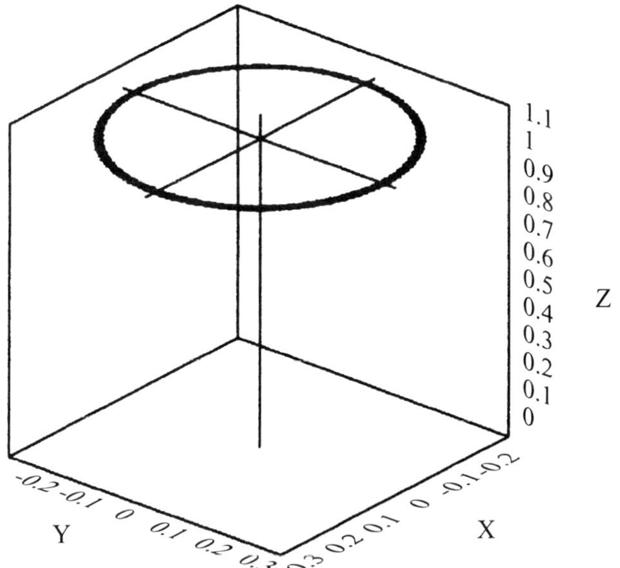

Figure 10.39.

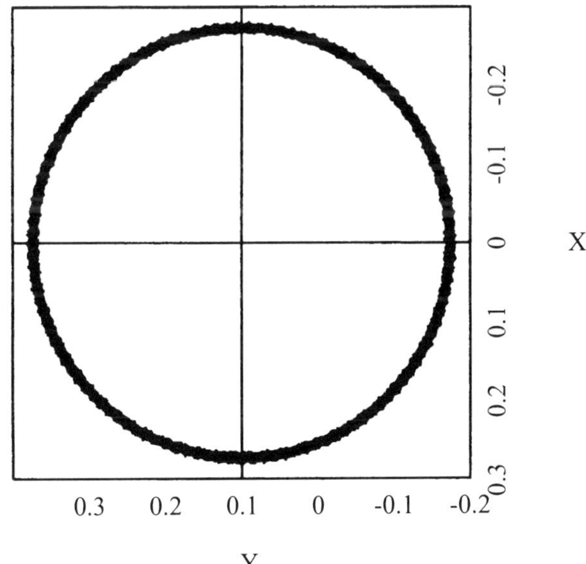

Figure 10.40.

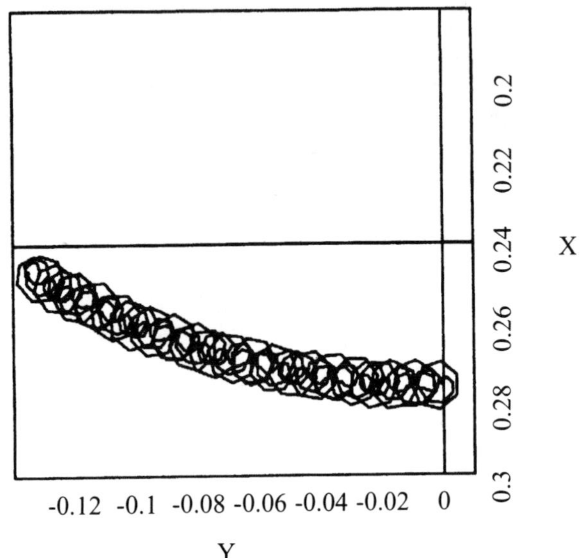

Figure 10.41.

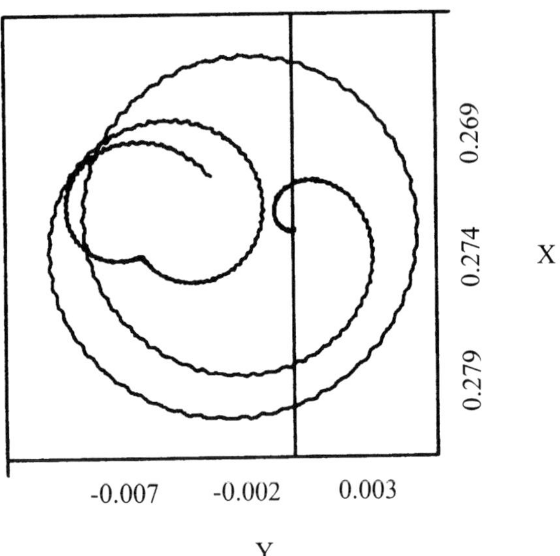

Figure 10.42.

Part III

Quantitative Modeling

11

Stress Wave Propagation in Slender Bars

11.1 Introduction

With the availability of today's advanced technology, experimental data are often available in various types and forms. It is important to examine as many sources of such data as is reasonable in formulating a particle model. In this chapter, we will show how to incorporate available stress and strain measurements for slender aluminum bars in order to simulate stress wave propagation in such bars. As an important byproduct of the development, we will see that the number of particles N in a particle model need not always be large in order to achieve excellent quantitative results. Indeed, we will require only $N = 20$. In addition, since N is relatively small, there is no substantial disadvantage in allowing square root routines into the leap frog calculations. So for variety, set the exponent parameters to $p = 2, q = 4$.

11.2 Force Formula Development

The basic ideas are summarized as follows. Only a one-dimensional array of particles will be considered. Each particle will interact locally only with its immediate neighbors. Experimental results will be incorporated into the local interaction formula, and the leap frog formulas will be applied with an exceptionally small time step, that is, Δt is chosen to be one half a microsecond, so that $\Delta t = 0.5 \times 10^{-6}$.

In order to allow for nonuniform mass distributions, we assume the local interaction formula

$$\vec{F}_{ij,k} = \left[-\frac{m_i m_j G}{(r_{ij,k})^p} + \frac{m_i m_j H}{(r_{ij,k})^q} \right] \frac{\vec{r}_{ji,k}}{r_{ij,k}}. \tag{11.1}$$

In the present chapter, because we will be guided by experimental data, we will, of necessity, have to deviate from cgs units. Thus, the units for (11.1) will not be prescribed until particular examples are discussed in Sect. 11.3. However, we will simplify (11.1) under the following assumptions. All N particles will be ordered linearly on an X-axis so that the particle numbers increase from left to right. Any particle P_i will be acted upon only by its adjacent particles. Thus, P_1 will be acted upon only by P_2, P_N will be acted upon only by P_{N-1}, and for $i = 2, 3, \ldots, N-1$, P_i will be acted upon by both P_{i-1} and P_{i+1}. Let us then consider the most complex case immediately, which occurs when P_i is an interior particle.

Let P_i be located at $x_{i,k}$. Assume first that $P_j = P_{i+1}$, which is located at $x_{i+1,k}$. Then the force on P_i due to P_{i+1} is

$$\vec{F}_{ij,k} = \left[-\frac{m_i m_j G}{(r_{ij,k})^p} + \frac{m_i m_j H}{(r_{ij,k})^q} \right] \frac{x_i - x_{i+1}}{x_{i+1} - x_i}. \tag{11.2}$$

Thus,

$$F_{ij,k} = \left[\frac{m_i m_j G}{(r_{ij,k})^p} - \frac{m_i m_j H}{(r_{ij,k})^q} \right], \quad j = i + 1. \tag{11.3}$$

Of course, the equilibrium distance r_0 for (11.3) satisfies

$$r_0^{p-q} = \frac{G}{H}. \tag{11.4}$$

Relation (11.4) establishes one constraint on the four parameters p, q, G, H. We next establish a second condition by introducing Young's modulus E. The strain $\epsilon_{ij,k}$ on P_i due to P_{i+1} is defined by

$$\epsilon_{ij,k} = \frac{r_{ij,k} - r_0}{r_0}, \quad j = i + 1. \tag{11.5}$$

The stress on P_i due to P_{i+1} is defined as $F_{ij,k}/A$, where $j = i + 1$ and A is the area over which the force acts, i.e., the cross-sectional area defined for P_i. The modulus of elasticity E is defined as the derivative of the stress with respect to the strain at the zero strain point. Hence,

$$E = \frac{\partial(F_{ij,k}/A)}{\partial \epsilon_{ij,k}} \bigg|_{\epsilon_{ij,k}=0} = \left[\frac{\partial(F_{ij,k}/A)}{\partial r_{ij,k}} \cdot \frac{\partial r_{ij,k}}{\partial \epsilon_{ij,k}} \right] \bigg|_{\epsilon=0} \tag{11.6}$$

$$= \left[\frac{\partial(F_{ij,k}/A)}{\partial r_{ij,k}} \bigg/ \frac{\partial \epsilon_{ij,k}}{\partial r_{ij,k}} \right] \bigg|_{\epsilon=0}.$$

But,

$$\frac{\partial(F_{ij,k}/A)}{\partial r_{ij,k}} = \frac{m_i m_j}{A} \left[\frac{-pG}{(r_{ij,k})^{p+1}} + \frac{qH}{(r_{ij,k})^{q+1}} \right] \tag{11.7}$$

and

$$\frac{\partial \epsilon_{ij,k}}{\partial r_{ij,k}} = \frac{1}{r_0}. \tag{11.8}$$

Hence, (11.4)–(11.8) imply that

$$G = \frac{E A r_0^p}{m_i m_j (q - p)}, \quad H = \frac{E A r_0^q}{m_i m_j (q - p)}, \tag{11.9}$$

so that substitution into (11.3) implies

$$\frac{F_{ij,k}}{A} = \frac{E}{q - p} \left[\left(\frac{r_0}{r_{ij,k}} \right)^p - \left(\frac{r_0}{r_{ij,k}} \right)^q \right]. \tag{11.10}$$

Thus, the force acting on P_i due to P_{i+1} is expressed in (11.10) as a function of $r_{ij,k}$; the constants E, A, r_0; and the parameters p and q. The relationship (11.10) represents the stress as a function of the strain through $r_{ij,k}$ and is called a stress–strain function.

Now consider the case where $P_j = P_{i-1}$, that is, P_j is the point to the left of P_i. In this case (11.3) is replaced by

$$F_{ij,k} = \left[-\frac{m_i m_j G}{(r_{ij,k})^p} + \frac{m_i m_j H}{(r_{ij,k})^q} \right], \quad j = i - 1.$$

But, the strain on P_i due to P_{i-1} is defined by

$$\epsilon_{ij,k} = \frac{r_0 - r_{ij,k}}{r_0}, \quad j = i - 1.$$

Hence, the derivation from (11.6) onward is the same because of the two sign changes, so that (11.10) is also valid in this case.

Of course, we are assuming that the force effect on P_i due to any point different from P_{i+1} and P_{i-1} is zero.

Now we set $p = 2, q = 4$ throughout this chapter.

11.3 Particle Model of a Slender Bar

A single chain of particles, as shown in Fig. 11.1, each linked with its immediate neighbors, has proved a satisfactory representation for the slender bars analyzed in this chapter. The particles were initially spaced at their equilibrium distance to avoid start-up transients. The mass associated with each element represents the distributed mass of the length of the bar represented by the elements. For simplicity, the stress and strain at a fixed right end were assumed to have the same values as the element to the left. Since gravity and bar support forces do not affect the axial stress

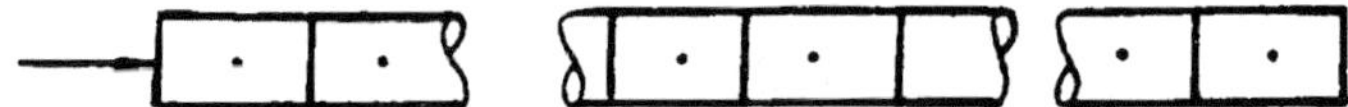

Figure 11.1 — Lumping of a distributed mass.

wave propagation, they were neglected. An impulsive force was applied to the first element at the left end of the bar as a compressive force along the axis. Except for bars with fixed end conditions, no axial restraint was considered in the analysis.

11.4 Examples

We now consider a variety of types of bars which are of engineering interest.

Example 11.1   Uniform bar with free-free ends
The simplest case for study of stress wave propagation is the constant cross-section bar with uniform, homogeneous density and elasticity. This case was analyzed for a one-half inch diameter aluminum bar. A half sine wave shaped force pulse of 35 microseconds duration and 5000 pounds peak magnitude was applied to one end of the bar. Both ends were otherwise unrestrained. The bar was 10 inches in length and was simulated by subdivision into 20 equal particles, each representing a one-half inch segment. Bar characteristics were chosen to facilitate comparison with results of a similar study in the literature (Sandlin (1970)). The mass of each segment is 0.2541×10^{-4} lb.sec^2/inch.

The bar strain response to the impulse is shown in Fig. 11.2 for several points along the span. The wave shapes and magnitudes are identical with those reported by Sandlin (1970).

In the case of a bar with a free end condition, a compression wave is reflected at the free end as a tension wave traveling in the opposite direction. According to one-dimensional wave theory the time between successive compression peaks is given by:

$$t = \frac{2L}{\sqrt{E/\rho}}$$

$$= \frac{2 \times 10}{\sqrt{10.6 \times 10^6 (386.4/0.1)}}$$

$$= 98.8 \text{ microseconds}$$

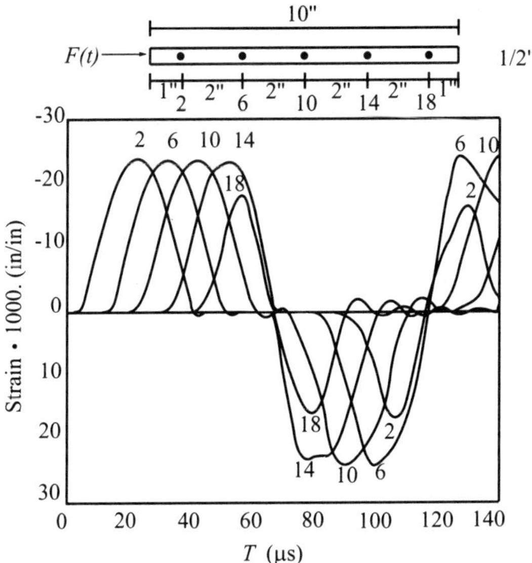

Figure 11.2 — Uniform bar strain history. Free-free end conditions.

and the peak strain magnitude is

$$\epsilon_{peak} = \frac{5000}{AE}$$

$$= \frac{5000}{(\pi/4)(1/2)^2 \times 10.6 \times 10^6}$$

$$= 2.402 \times 10^{-3} \text{ inches/inch.}$$

Examining the strain history of point 6 in Fig. 11.2, we see that the two compression peaks are approximately 97 microseconds apart and that their strain magnitudes are near 2.40×10^{-3} inches per inch. Thus, the numerical results show good agreement with the theoretical predictions.

Example 11.2   Uniform bar with free-fixed ends

The uniform cylindrical bar was also analyzed for the case where the end opposite the applied impulse was fixed. This case was simulated mathematically by specifying the position of the end particle to remain fixed at its initial position. Otherwise, properties of the bar and the applied impulse were the same as for the free-free case.

Response of the free-fixed bar is shown in Fig. 11.3. The compression wave in this case is reflected at the fixed end as another compression wave traveling in the

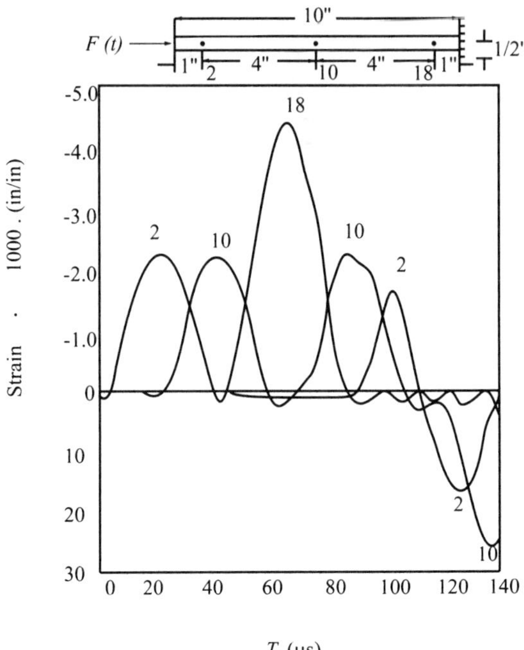

Figure 11.3 — Uniform bar strain history. Free-fixed end conditions.

opposite direction. Again, the results of the analysis agree identically with those in the literature (Sandlin (1970)).

Example 11.3   Free-free bar with stepped area change

A free-free cylindrical bar with a stepped area change at its midspan was analyzed to compare the predicted wave propagation characteristics with those described elsewhere by Sandlin (1970). Bars of this type are commonly used as shafting in machinery with gears, pulleys or sprockets. The pulse input and the bar dimensions of the left half of the bar are the same as those of the uniform bar discussed in the previous examples. The cross-section area of the right half of the bar was twice that of the left half. The strain time history for the stepped bar is shown in Fig. 11.4. The results agree closely with those found by others and the "bump" noted by Sandlin (1970), produced by reflection from the step, is evident.

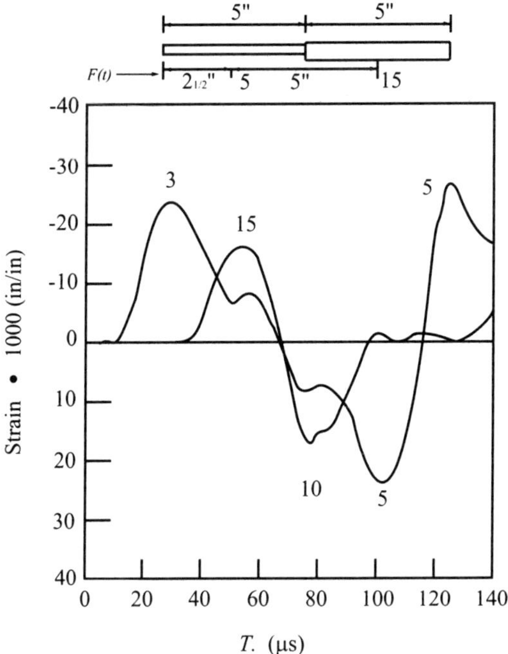

Figure 11.4 — Stepped area bar strain history. Free-free end conditions.

12

Colliding Microdrops of Water

12.1 Introduction

Collision of microdrops are important in microwave, chemical nucleation and raindrop studies (Adam, Lindblad and Hendricks (1968), Peterson (1985), Simpson and Haller (1988)). In this chapter, we shall show how to simulate collisions of microdrops of water. Since the interaction during collisions will be independent of gravity, we will neglect long-range forces.

12.2 Mathematical and Physical Considerations

An elementary water molecule potential (Hirschfelder, Curtiss and Bird (1965)) is:

$$\phi(r) = 1.9646833 \times 10^{-13} \left[\left(\frac{2.725}{r} \right)^{12} - \left(\frac{2.725}{r} \right)^{6} \right] \quad erg \qquad (12.1)$$

in which r is measured in angstroms. From (12.1) the force $\vec{F}$, in dynes, between two molecules r Å apart has magnitude F given by

$$F(r) = 4.325809 \times 10^{-5} \left[2 \left(\frac{2.725}{r} \right)^{13} - \left(\frac{2.725}{r} \right)^{7} \right]. \qquad (12.2)$$

In considering how to proceed in the present chapter, let us recall that all mathematical models are only approximations of the real thing. For this reason, we will not use (12.2), but will develop a modified, simpler formula. Direct use of a molecular formula will be explored in Chapter 16.

Consider then a least square fit of (12.2) by the function

$$F^*(r) = 4.325809 \times 10^{-5} \left[-\frac{G}{r^3} + \frac{H}{r^5} \right]. \tag{12.3}$$

For the fit one can determine as many data points $(r, F(r))$ as one desires from (12.2). For simplicity, let us consider only the five values $r = 2.5, 2.75, 3.0, 3.25, 3.5$, which straddle the equilibrium point $r = 3.06$. Then, from (12.2) with $C = [4.325809]^{-1} 10^5$,

$$CF(3.5) = -0.09616 \,, \, CF(3.25) = -0.08888 \,, \tag{12.4}$$

$$CF(3.0) = 0.06291 \,, \, CF(2.75) = 0.83084 \,, \, CF(2.5) = 4.30357 \,,$$

from which it follows that the least square fit is

$$F^*(r) = 4.325809 \times 10^{-5} \left[-\frac{115}{r^3} + \frac{1104}{r^5} \right]. \tag{12.5}$$

Since $\vec{F}$ is in dynes and since the mass of water molecule is $30.103 \times 10^{-24} g$, a dynamical equation which describes the motion of one water molecule which interacts with only one other water molecule r Å away is

$$30.103 \times 10^{-24} \vec{a} = 4.325809 \times 10^{-5} \left[-\frac{115}{r^3} + \frac{1104}{r^5} \right] \tag{12.6}$$

Changing to Å/s^2 and also introducing the computationally convenient time transformation $T = 10^{13.5} t$ yields the dynamical equation

$$\frac{d^2 \vec{r}}{dT^2} = -\frac{16.5}{r^3} + \frac{158.6}{r^5}. \tag{12.7}$$

For a system of N water molecules $P_1, P_2, \ldots, P_N$, it follows from equation (12.7) that, from given initial data, the motion of each P_i can be determined by solving the system of second order, nonlinear, ordinary differential equations

$$\frac{d^2 \vec{r}_i}{dT^2} = \sum_{\substack{j=1 \\ j \neq i}}^{N} \left(-\frac{16.5}{r_{ij}^3} + \frac{158.6}{r_{ij}^5} \right) \frac{\vec{r}_{ji}}{r_{ij}}, \tag{12.8}$$

in which $\vec{r}_i$ is the position vector of P_i, $\vec{r}_{ji}$ is the vector from P_j to P_i and r_{ij} is the magnitude of $\vec{r}_{ji}$.

12.3 Examples

Before studying the interaction of two water drops, it is necessary to generate a single drop, which is done as follows, and for physical reasons, we will do this first in two dimensions. Since $\phi(2.725) = 0$, let us consider, for variety, a regular triangular mosaic of points (x_i, y_i), $i = 1, 2, \ldots, 9000$, given by:

$$x_1 = -68.125, \ y_1 = -58.997975, \ x_{52} = -66.7625, \ y_{52} = -56.638056,$$

$$x_{i+1} = 2.725 + x_i, \ y_{i+1} = y_1, i = 1, 2, \ldots, 50,$$

$$x_{i+1} = 2.725 + x_i, \ y_{i+1} = y_{52}, i = 52, 53, \ldots, 100,$$

$$x_i = x_{i-101}, \ y_i = 4.719384 + y_{i-101}, i = 102, 103, \ldots, 9000.$$

From these we choose only those which satisfy $x_i^2 + y_i^2 \leq 2320$, thus yielding 1128 points which lie in a relatively circular pattern. At each such point (x_i, y_i) we place a water molecule P_i.

Each of the 1128 water molecules P_i is now allowed to interact with all other molecules in accordance with equation (12.8). For simplicity, we assume first that all initial velocities are zero. The leap frog formulas are then applied numerically with $\Delta T = 0.0002$ until $T = 11.2$. At this time the system has contracted maximally, so that its energy should be almost all potential. Thus, at $T = 11.2$ all velocities are reset to zero and the system is allowed to interact until $T = 14.0$, at which time all velocities are again reset to zero. Thereafter, the molecules are allowed to interact without further damping. The resulting system configurations are shown at $T = 14.0, 16.8, 19.6, 22.4, 25.2, 28.0, 30.8$ and 33.6 in Figs. 12.1(a)–(h), respectively. These figures show the presence of surface waves, which, in fact, are due to the system's contractions and expansions with time. Note also that the density at any time is always greater in the interior of the system than at the boundary, which is consistent with the *surface tension theory* which holds that surface molecules are in an attraction mode.

Our real interest, however, is in three dimensions, not two. The discussion has been limited thus far to two because it is easier in this case to demonstrate pictorially that molecular fluid models contain surface tension inherently. There is no need, as when one considers the Navier-Stokes equations, to impose surface tension on the model.

In three dimensions (Greenspan and Heath (1991)), then consider $N = 4102$ water molecules $P_1, P_2, \ldots, P_N$ which interact in accordance with the dynamical system

$$\frac{d^2\vec{r}_i}{dT^2} = \sum_{\substack{j=1 \\ j\neq i}}^{N} \left(-\frac{16.5}{r_{ij}^3} + \frac{158.6}{r_{ij}^5} \right) \frac{\vec{r}_{ji}}{r_{ij}}, \quad i = 1, 2, \ldots, N \tag{12.9}$$

Our objective will be to study collision modes of two water drops, so for convenience, a single water drop will be generated first.

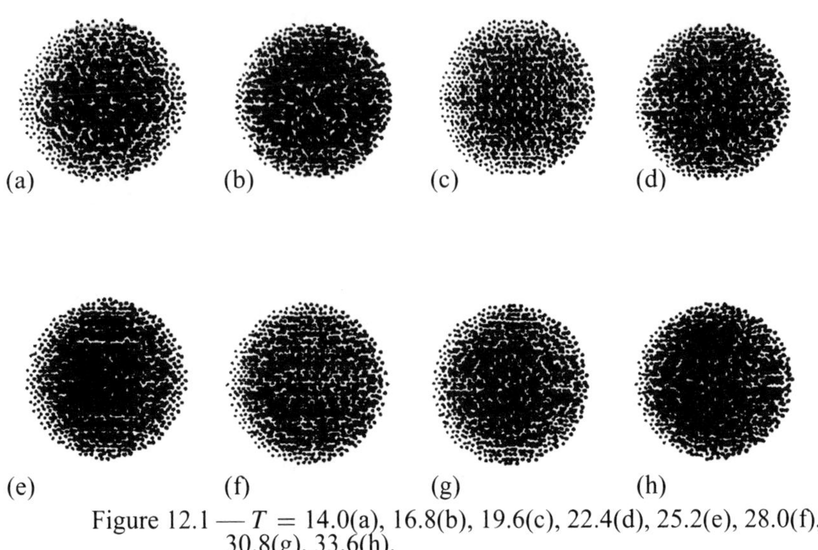

Figure 12.1 — $T = 14.0$(a), 16.8(b), 19.6(c), 22.4(d), 25.2(e), 28.0(f), 30.8(g), 33.6(h).

Consider that portion of three space for which $-31 \leq x \leq 31$, $-31 \leq y \leq 31$, $-31 \leq z \leq 31$, and let water molecules be placed at the grid points which result for $\Delta x = \Delta y = \Delta z = 3.1$. This time, again for variety, we have chosen the value $r = 3.1$ which makes F equal to zero, rather than the value which makes ϕ equal to zero. (Again, however, we will have to impose a damping procedure, which indicates that neither initial choice of r seems to be the superior one.)

Next, molecules outside the sphere whose equation is $x^2 + y^2 + z^2 = 26^2$ are deleted and each of the remaining molecules is assigned a random velocity in the range $|v| \leq 0.02$. At initial time, then, there are 2517 molecules which are thereafter allowed to interact in accordance with (12.9) for 31,000 time steps. At T_{31000}, all molecules whose position coordinates satisfy $r > 26$ are deleted, reducing the number to 2051. The simulation is then continued to T_{88000}, but with all velocities reset to zero at T_{40500}, T_{48000}, T_{53000}, T_{58000}, T_{63000}, T_{68000} and T_{73000}. At T_{78000} the velocities are damped by the factor 0.5. The damping process so imposed cools the molecular configuration so that at T_{88000} the temperature (Hirschfelder, Curtiss and Bird (1965)) of the resulting drop is $45°$C. This drop is shown in Fig. 12.2.

In order to study collision modes, the single drop generated above is duplicated by mirror imaging. The resulting two drops are set symmetrically 3 Å apart about the YZ plane, as shown in Fig. 12.3. To elucidate the motions of individual molecules during collision, the drops are displayed in different shades. To avoid complete symmetry, the velocity of any molecule and that of its mirror image molecule are taken to be the same. In addition, the time counter is reset to zero.

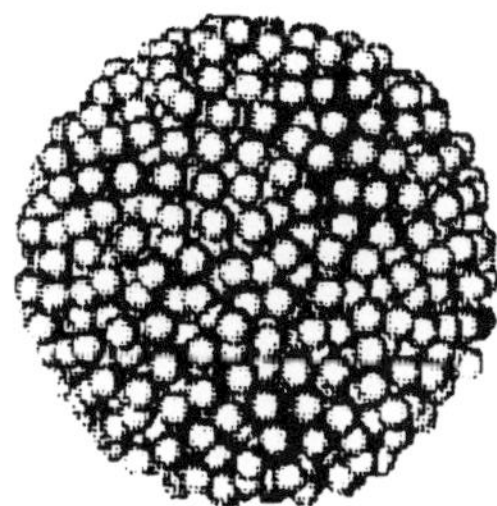

Figure 12.2 — A microdrop of water at $45°C$.

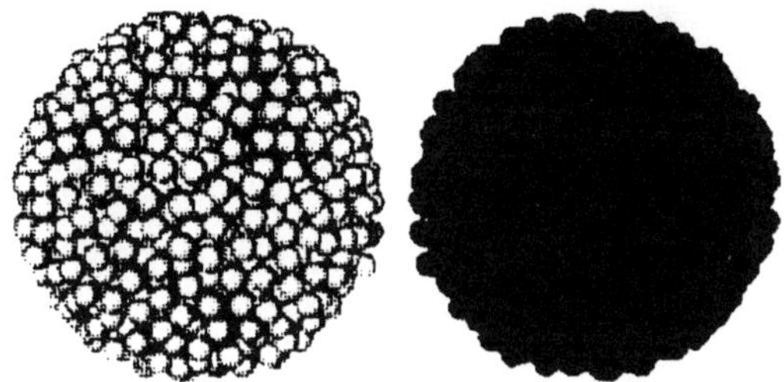

Figure 12.3 — Two microdrops of water 3 Å apart.

To simulate collision, we will assume that each molecule of the light drop, on the left in Fig. 12.3, has its velocity increased initially by $\vec{v}^*$, while each molecule of the dark drop has its velocity decreased by $\vec{v}^*$.

As a first case, let $\vec{v}^* = (0, 0, 0)$, so that the two drops are allowed to interact with no changes in velocity. Then, Fig. 12.4 shows, at the indicated times, an oblate spheroid oscillation mode. After an extended period of time, the large boundary gradients due to surface tension transform this mode into a relatively spherical drop which exhibits small oscillations throughout its surface. Indeed, in this example and in all cohesive interactions to be described, the so-called oscillation modes are, in reality, dynamical configurations which, in time, transform into a spherical configuration.

Next, set $\vec{v}^* = (2.2, 0.2, 0)$. Figure 12.5 then shows, at the indicated times, the development of a raindrop mode (Peterson (1985)).

Setting $\vec{v}^* = (2.0, 4.5, 0)$ yields, as shown in Fig. 12.6, a dumbbell mode (Adam, Linblad and Hendricks (1968), Simpson and Haller (1988))

Next, selecting the largest speed of any case yet considered by the choice $\vec{v}^* = (0.2, 8.0, 0)$ yields, as shown in Fig. 12.7, a noncohesive, brush-type collision in which each of the drops forms a teardrop mode (Simpson and Haller (1988)).

Finally, increasing the speed still further, but in a fashion which results in more direct collision, the choice $\vec{v}^* = (5.0, 10.0, 0)$ yields, as shown in Fig. 12.8,

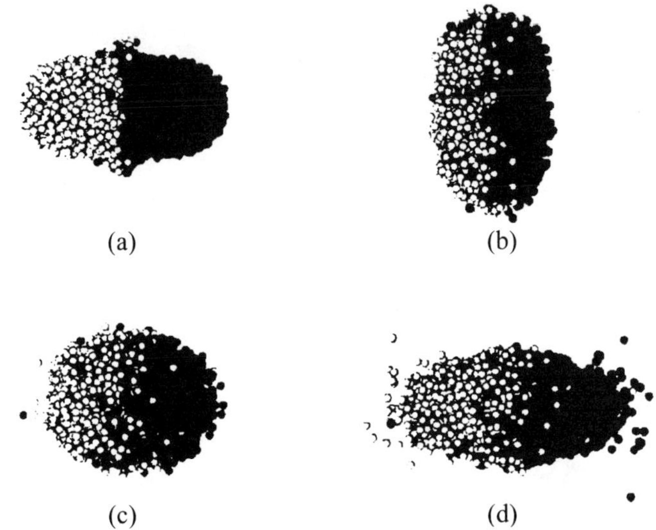

Figure 12.4 — Oscillating oblateness mode. (a) T_{9500}, (b) T_{8500}, (c) T_{27500}, (d) T_{36500}.

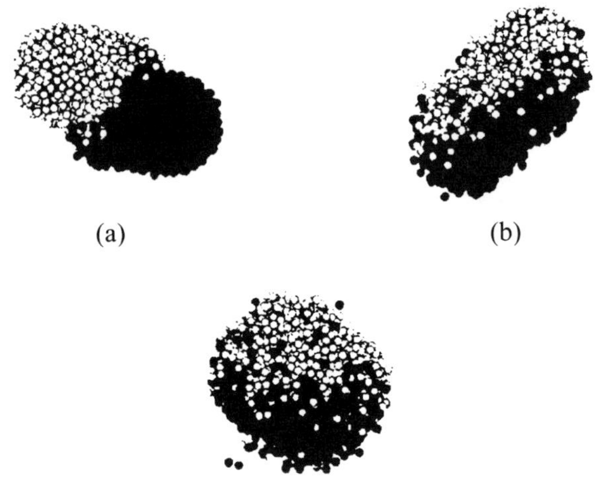

Figure 12.5 — Raindrop mode. (a) T_{9500}, (b) T_{18500}, (c) T_{27500}.

a noncohesive collision which exhibits an initial clean slicing effect and the molecular transfer during and after separation.

Of the additional cases considered, we found that choices of $\vec{v}^*$ which yielded high speeds resulted in explosive type reactions. Thus, for $\vec{v}^* = (12.0, 0.2, 0)$,

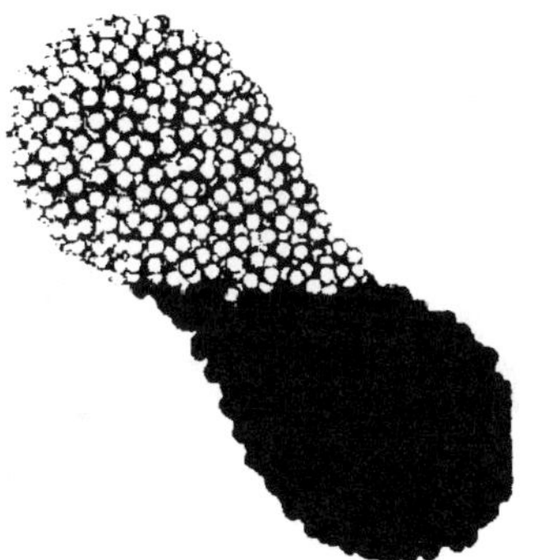

Figure 12.6 — Dumbbell mode.

Figure 12.7 — Brush-type collision with teardrop modes.

this type of reaction is shown in Fig. 12.9. Large momentum effects, as in the case $\vec{v}^* = (0.2, 5.0, 0)$, would often have dumbbell modes develop into a peanut shape in their transition to sphericity, as shown in Fig. 12.10.

Small perturbations of each collision mode described above yielded entirely similar results confirming the physical stability of the nodes.

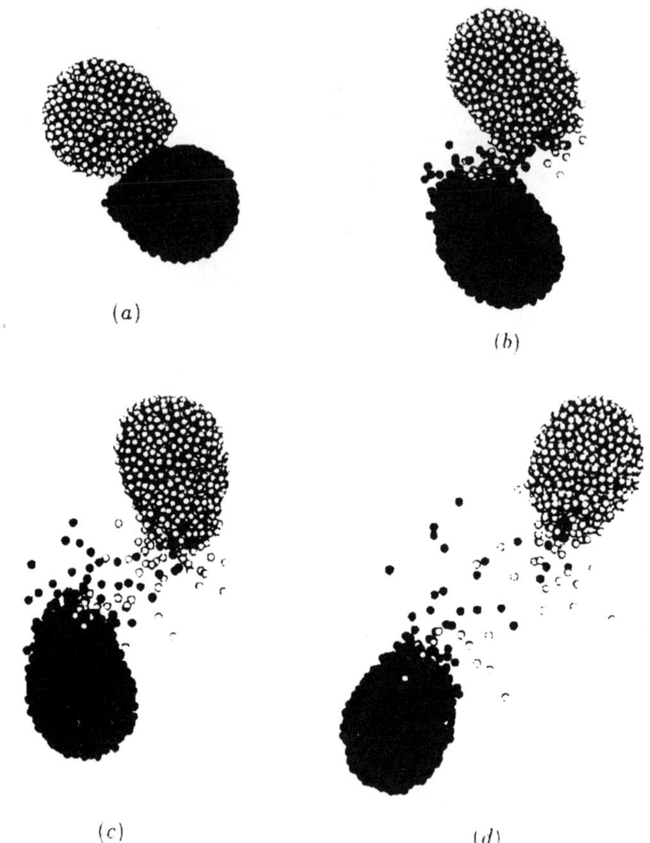

Figure 12.8 — Soft collision. (a) T_{4000}, (b) T_{9000}, (c) T_{13500}, (d) T_{18000}.

Figure 12.9 — Direct hard collision.

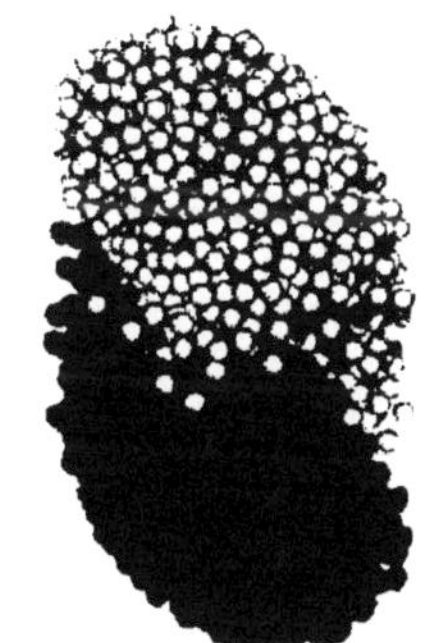

Figure 12.10 — Peanut mode.

13

Crack Development in a Stressed Copper Plate

13.1 Introduction

The problem we will consider in this chapter is that of crack development in a stressed plate. The ability to predict where a crack will develop in a stressed plate is of fundamental importance in the design of buildings, aircraft and nuclear reactors. Specifically, we will simulate crack and fracture development in a stressed, slotted copper plate.

13.2 Formula Derivation

An approximate potential function for the interaction of two copper atoms r Å apart is (Greenspan (1989 b))

$$\phi(r) = -\frac{1.398068 \times 10^{-10}}{r^6} + \frac{1.55104 \times 10^{-8}}{r^{12}}\,\text{erg}. \qquad (13.1)$$

From (13.1) it follows that the magnitude F of the force $\vec{F}$, in dynes, between two copper atoms r Å apart is

$$F(r) = -\frac{8.388408 \times 10^{-2}}{r^7} + \frac{1.861248 \times 10}{r^{13}}. \qquad (13.2)$$

The minimum ϕ occurs when $F(r) = 0$, that is, at $r = 2.46$ Å and yields

$$\phi(2.46) = -3.15045 \times 10^{-13}\,\text{erg}. \qquad (13.3)$$

With these observations made, let us then consider a rectangular copper plate which is approximately 8 cm $\times$ 11.4 cm. To simulate the plate, let the points P_i

with respective coordinates (x_i, y_i), $i = 1, 2, \ldots, 2713$, be defined by

$$x(1) = -3.9, \quad y(1) = -5.71576764,$$

$$x(41) = -4.0, \quad y(41) = -5.54256256,$$

$$x(i + 1) = x(i) + 0.2, \quad y(i + 1) = y(1), \quad i = 1, 2, \ldots, 39$$

$$x(i + 1) = x(i) + 0.2, \, y(i + 1) = y(41), \, i = 41, 42, \ldots, 80$$

$$x(i) = x(i - 81), \, y(i) = y(i - 81) + 2(0.17320508),$$

$$i = 82, 83, \ldots, 2713.$$

The resulting arrangement is shown in Fig. 13.1. The (x_i, y_i) are vertices of a regular triangular mosaic in which the distance from any P_i to an immediate neighbor is 0.2 cm. The P_i are assumed to represent particles, or aggregates of molecules, of an 8 cm $\times$ 11.43 cm rectangular copper plate. The neighbors of any P_i are those particles which are 0.2 cm from P_i. The neighbors of any P_i are defined to be the neighbors of P_i for all time.

In order to determine a mass m for each P_i, we use total mass conservation. Suppose the rectangular plate were to be filled with copper atoms using, again, a regular triangular mosaic, but in which the distance between two immediate neighbors is 2.46 Å. Then the number N^* of atoms in the plate is approximately,

$$N^* = \frac{8 \times 10^8}{2.46} \times \frac{11.43 \times 10^8}{2.13} = 1.745 \times 10^{17}. \tag{13.4}$$

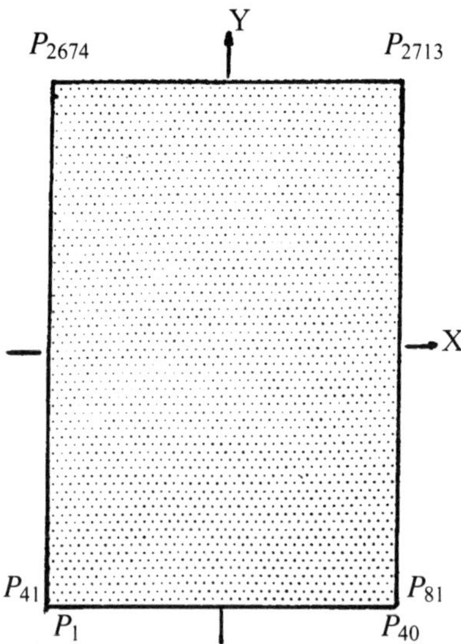

Figure 13.1 — The initial configuration.

Since the mass of a copper atom is $1.0542 \times 10^{-22}g$, the total mass M of these copper atoms is then $M = 1.840 \times 10^{-5}g$. Distributing the mass over 2713 particles yields a particle mass m given by

$$m = 6.782 \times 10^{-9}g \tag{13.5}$$

To determine computationally convenient force and potential formulas, we utilize energy conservation. Since the minimum potential between two copper atoms is given by equation (13.3), it follows under the assumption of zero kinetic energy, as in ideal crystal theory, that the total energy E^* of the system of atoms is, approximately,

$$E^* = 3(1.745)10^{17}(-3.15045)10^{-13} = -1.6493 \times 10^5 \text{ erg.} \tag{13.6}$$

Let us assume now that the force $\vec{F}$, in dynes, between two particles has magnitude F given by

$$F(R) = -\frac{G}{R^3} + \frac{H}{R^5} \tag{13.7}$$

in which R is measured in centimeters. Hence

$$\phi(R) = -\frac{0.5G}{R^2} + \frac{0.25H}{R^4} \text{ erg.} \tag{13.8}$$

Assuming $\phi(R)$ is minimal for $R = 0.2$, so that $F(0.2) = 0$, implies

$$-\frac{G}{(0.2)^3} + \frac{H}{(0.2)^5} = 0 . \tag{13.9}$$

Approximating the total energy E of the particle system in the fashion used to obtain E^* yields

$$E = 3(2713)\left(-\frac{G}{2(0.2)^2} + \frac{H}{4(0.2)^4}\right) \text{ erg.} \tag{13.10}$$

Equating E and E^* implies

$$-\frac{G}{2(0.2)^2} + \frac{H}{4(0.2)^4} = -20.264. \tag{13.11}$$

The solution of equations (13.9) and (13.11) is $G = 3.24224$, $H = 0.12969$, so that equation (13.7) takes the specific form

$$F(R) = \frac{-3.24224}{R^3} + \frac{0.12969}{R^5} . \tag{13.12}$$

Now, the force between two particles must be local in the presence of gravity. Thus, we will introduce a normalizing constant a such that at a distance 0.34 cm the force between two particles is small relative to gravity. This is essential because we have assumed that forces act only between neighbors, and initially the distance from any P_i to a neighbor is 0.2 cm. If we can define "small relative to gravity" to mean 0.1% of the effect of gravity, and assume that

$$a| - 3.24224/(0.34)^3 + 0.12969/(0.34)^5| < (0.001)980m \qquad (13.13)$$

then $a \sim (1.25)10^{-10}$.

Since the effect of gravity will be relatively insignificant in the problems to be considered, the dynamical equation for the motion of each particle is then

$$m\frac{d^2 \vec{R}_i}{dt^2} = 1.25 \times 10^{-10} \sum \left[\left(\frac{-3.24224}{(R_{ij})^3} + \frac{0.12969}{(R_{ij})^5} \right) \frac{\vec{R}_{ji}}{R_{ij}} \right] \qquad (13.14)$$

in which $\vec{R}_{ji}$ is the vector P_j to P_i and summation is taken over the neighbors of P_i. From equation (13.5) and introducing the computationally convenient transformations $R^* = 4R, T^2 = 10t^2$, equation (13.14) reduces to

$$\frac{d^2 \vec{R}_i^*}{dT^2} = \sum \left[\left(-\frac{1.52981}{(R_{ij}^*)^3} + \frac{0.97908}{(R_{ij}^*)^5} \right) \frac{\vec{R}_{ji}^*}{R_{ij}^*} \right]. \qquad (13.15)$$

For the distance D^* at which a particle bond breaks we choose, as recommended by Ashurst and Hoover (1976), the value R^* at which dF/dR^* first becomes negative. From equation (13.15), then, $D^* = 1.033$.

13.3 Examples

As a first example, consider a slotted plate, that is, as shown in Fig. 13.2, a plate in which the 15 particles P_i, $i = 1070 + 41k$, $k = 0, 1, 2, \ldots, 14$, have been removed. At each time step the particles in the bottom row are relocated so that their Y coordinates are decreased by 0.00002 units. But also the particles in the top row have their Y coordinates increased by 0.00002 units per time step. Thereby, the plate is stretched. Solving system (13.15) numerically with leap frog formulas with time step $\Delta T = 0.0001$ yields the following results. Figures 13.3–13.5 show the developing force field throughout the bottom half of the plate at the times $T = 2.0, 6.0, 10.0$. In these figures the force field is represented by vectors emanating from the centers of the particles. Figures 13.3–13.5 reveal the stress effect being transmitted to the interior of the plate. The fact that this transmission is not instantaneous serves to differentiate a nonimpulsive force from an impulsive one, which would simply tear the top and bottom of the plate. Figure 13.5 reveals large forces on the left and right sides of the slot, which yield a widening of the slot. Figure 13.6, at $T = 12.8$, shows clearly from the force field that the first crack occurs at the lower left of the slot, and hence by symmetry, simultaneously at the upper right. Figures 13.7–13.9 show the gross effect on the plate at the respective times $T = 13.8, 15.8, 18.8$. Figure 13.10 shows the associated row dislocation at $T = 16.8$.

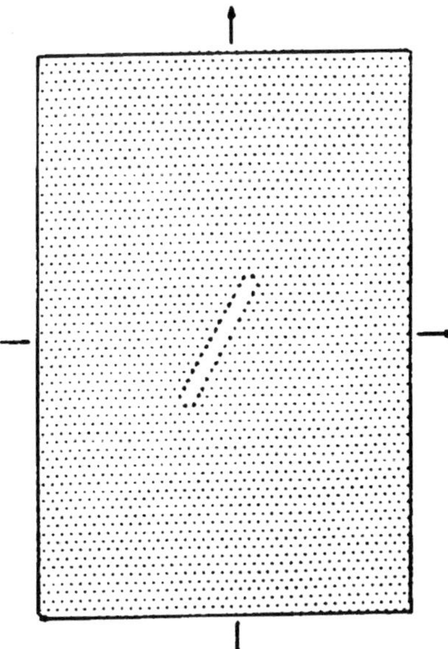

Figure 13.2 — The slotted plate.

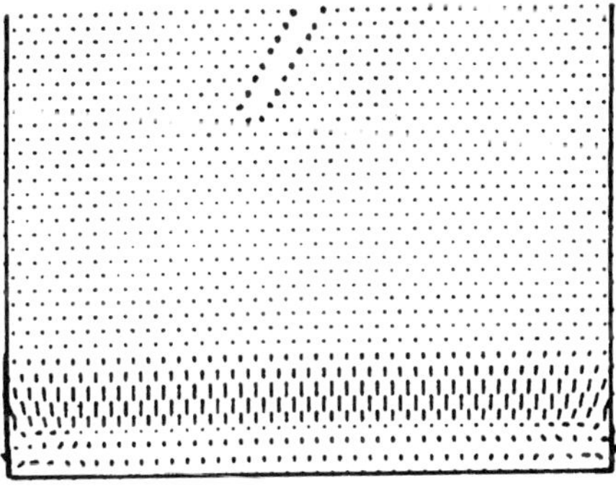

Figure 13.3 — $T = 2.0$.

Figures 13.11 and 13.12 show the fracture of a full plate under shear at the times $T = 20.0$ and $T = 30.0$. In this case, the top and bottom rows were again moved 0.00002 units per time step in the Y direction, while they were moved simultaneously, and again symmetrically with respect to the origin, 0.000005 units in the X direction per time step.

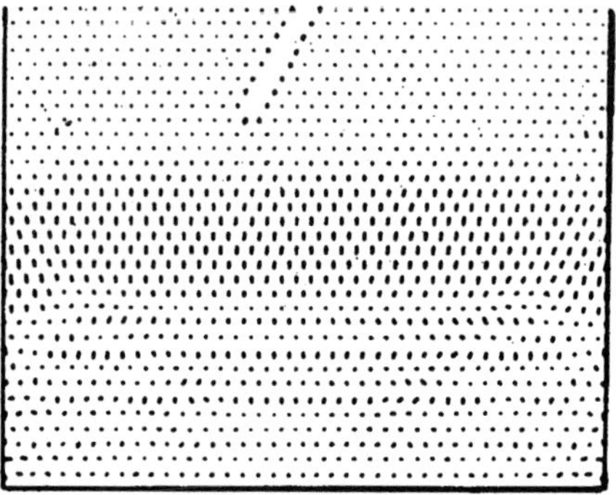

Figure 13.4 — $T = 6.0$.

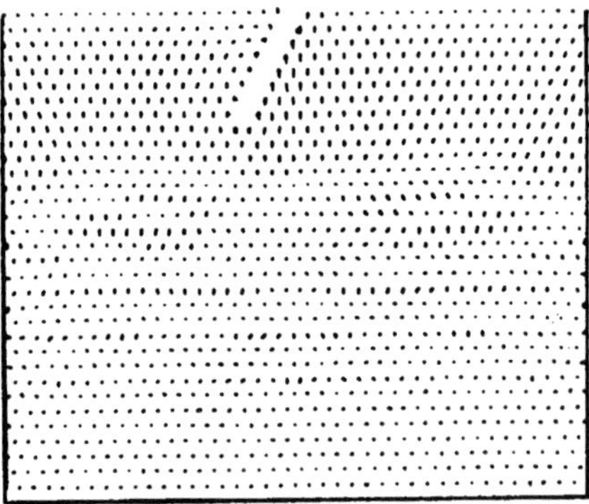

Figure 13.5 — $T = 10.0$.

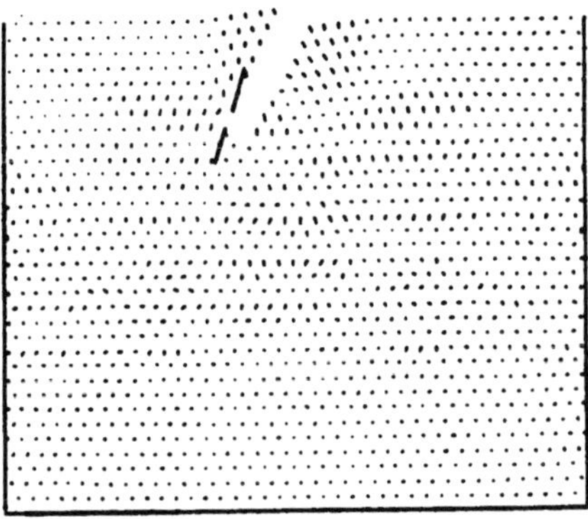

Figure 13.6 — $T = 12.8$.

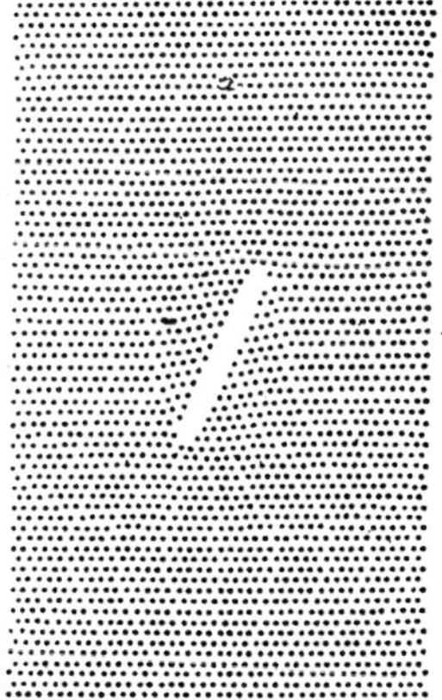

Figure 13.7 — $T = 13.8$.

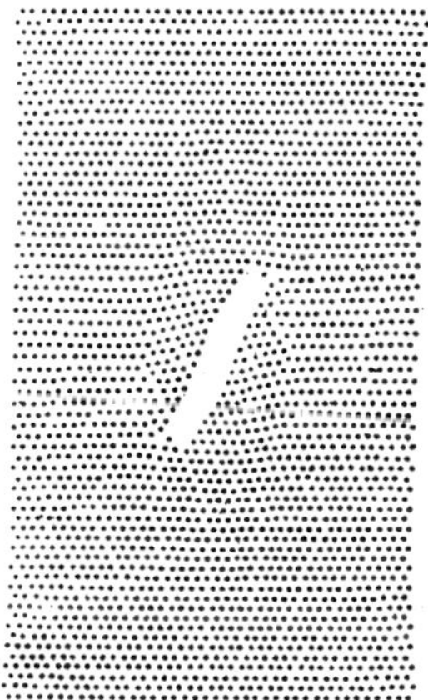

Figure 13.8 — $T = 15.8$.

Figure 13.9 — $T = 18.8$.

Figure 13.10 — Row dislocation at $T = 16.8$.

Figure 13.11 — $T = 20.0$.

Figure 13.12 — $T = 30.0$.

14

Liquid Drop Formation on a Solid Surface

14.1 Introduction

In this chapter, we will show how to simulate the formation of a liquid drop, taken to be water, on a horizontal solid surface, taken to be graphite. At present, for computational simplicity, attention will be restricted to two-dimensional simulations. Nevertheless, all the ideas and methods do extend to three dimensions.

14.2 Local Force Formulas

Consider the system of 823 water particles P_1, P_2, ..., P_{823} shown in Fig. 14.1. These particles are arranged on a regular triangular mosaic but in a relatively circular pattern. The edge length of each triangle in the mosaic is 0.0305871 cm, the rationale of which will be explained shortly. The algorithm for generating the positions $\vec{r}(i) = (x(i), y(i))$ and velocity $\vec{v}(i) = (v_x(i), v_y(i))$ for each P_i is given as follows. From the 9500 points with

$$x(1) = -0.6117420,\ y(1) = -0.7946762,\ v_x(1) = 0.0,\ v_y(1) = 10^{-9},$$

$$x(42) = -0.5964485,\ y(42) = -0.7681870,\ v_x(42) = 10^{-9},\ v_y(42) = 0.0,$$

$$x(i+1) = x(i) + 0.0305871,\ y(i+1) = y(1),\ v_x(i+1) = 0.0,$$

$$v_y(i+1) = v_y(1),\quad i = 1, 2, 3, \ldots, 40,$$

$$x(i+1) = x(i) + 0.0305871,\ y(i+1) = y(42),\ v_x(i+1) = v_x(42),$$

$$v_y(i+1) = 0.0,\quad i = 42, 43, \ldots, 80,$$

$$x(i) = x(i - 81), \; y(i) = y(i - 81) + 0.0529784, \; v_x(i) = -v_x(i - 81),$$

$$v_y(i) = -v_y(i - 81), \quad i = 82, 83, \ldots, 9500,$$

choose only those which satisfy

$$[x(i)]^2 + [y(i)]^2 \le 0.2123 . \tag{14.1}$$

These are the 823 points shown in Fig. 14.1. Note that the horizontal radius of this relatively circular set contains 15 particles, so that the radius r is approximately

$$r = 15(0.0305871) = 0.4588065 \; \text{cm} . \tag{14.2}$$

For the interaction of the particles we assume a force $\vec{F}_1$, in dynes, between two particles R cm apart, with magnitude F_1 given

$$F_1(R) = -\frac{G}{R^3} + \frac{H}{R^5} . \tag{14.3}$$

From (14.3), then, in ergs,

$$\phi_1(R) = -\frac{G}{2R^2} + \frac{H}{4R^4} . \tag{14.4}$$

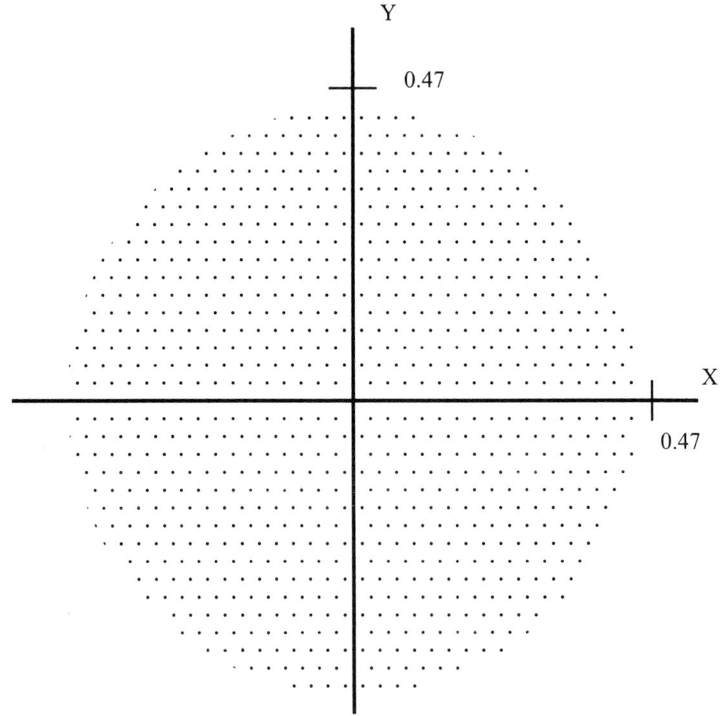

Figure 14.1.

For the system shown in Fig. 14.1 we assume that $F_1 = 0$ between two neighbors, so that

$$-(0.0305871)^2 G + H = 0 . \tag{14.5}$$

Assuming zero kinetic energy, the total energy E_1 of the particle system is, approximately,

$$E_1 = 3(823) \left(-\frac{G}{2(0.0305871)^2} + \frac{H}{4(0.0305871)^4} \right) . \tag{14.6}$$

Now, for actual water molecules we use the approximation (12.1), that is,

$$\phi(r) = (1.9646383)10^{-13} \left[\left(\frac{2.725}{r} \right)^{12} - \left(\frac{2.725}{r} \right)^{6} \right] \text{erg} , \tag{14.7}$$

in which the distance r between two molecules is measured in angstroms. From (14.7) the magnitude of the force $\vec{F}$, in dynes, is

$$F(r) = (1.9646383)10^{-5} \left(12 \frac{(2.725)^{12}}{r^{13}} - 6 \frac{(2.725)^6}{r^7} \right) . \tag{14.8}$$

Note from (14.8) that $F(r) = 0$ implies $r = 3.05871$ Å. (Our choice of 0.0305871 cm for the construction of Fig. 14.1 was based on the angstrom measurement for the purpose of simplifying later calculations.) We now fill the circle shown in Fig. 14.1 with molecules which are vertices of a regular triangular grid with edge length 3.05871 Å. The number N of molecules which fill the region is approximately

$$N = \pi \left(\frac{15(0.0305871)}{(3.05871)10^{-8}} \right)^2 \approx (706.858)10^{12} . \tag{14.9}$$

Assuming zero kinetic energy, the total energy E of the molecular system is approximately

$$E = 3(707)10^{12}(1.9646383)10^{-13} \left[\left(\frac{2.725}{3.05871} \right)^{12} - \left(\frac{2.725}{3.05871} \right)^6 \right] \text{erg} \tag{14.10}$$

or

$$E = -104.1745 .$$

Note that the expression in the square brackets of (14.10) is simply $\left[\left(\frac{1}{2} \right)^2 - \left(\frac{1}{2} \right) \right]$. Equating E_1 and E implies

$$-534.4332G + 285618.8H = -0.04219299 . \tag{14.11}$$

The solution of (14.5) and (14.11) is

$$G = (1.57898)10^{-4}, \qquad H = (1.47725)10^{-7} ,$$

so that (14.3) and (14.4) can be given explicitly as

$$F_1(R) = -\frac{1.57898}{R^3}10^{-4} + \frac{1.47725}{R^5}10^{-7}, \qquad (14.12)$$

$$\phi_1(R) = -\frac{7.8949}{R^2}10^{-5} + \frac{3.693125}{R^4}10^{-8}. \qquad (14.13)$$

Note that (14.13) can be rewritten in the form

$$\phi_1(R) = 4\epsilon_1\left[-\left(\frac{\sigma_1}{R}\right)^2 + \left(\frac{\sigma_1}{R}\right)^4\right], \qquad (14.14)$$

in which $\epsilon_1 = 0.0421929$ and $\sigma_1 = 0.0216284$.

Note also that since the mass of a single water molecule is approximately $m = (30.103)10^{-24}$g, distributing the total water molecule mass over the 823 particles yields an individual water particle mass M_1 given by

$$M_1 = (2.586)10^{-11}\text{g}. \qquad (14.15)$$

Using the very same line of reasoning as for water, let us proceed to develop appropriate formulas for graphite. A five row, 1003 particle slab of graphite particles is generated using a regular triangular mosaic with edge length 0.03834 cm by the formulas

$$x(1) = -3.834, \quad y(1) = 0.0, \quad x(202) = -3.81085, \quad y(202) = 0.03317,$$

$$x(i+1) = x(i) + 0.03834, \quad y(i+1) = y(1), \quad i = 1, 2, \ldots, 200,$$

$$x(i+1) = x(i) + 0.03834, \quad y(i+1) = y(202), \quad i = 202, 203, \ldots, 400,$$

$$x(i) = x(i-401), \quad y(i) = 0.06634 + y(i-401), \quad i = 402, 403, \ldots, 1003.$$

The resulting slab is 7.688 cm wide and 0.13268 cm high and is shown in Fig. 14.2. One should note immediately the difference in units between Figs. 14.1 and 14.2.

Let the force $\vec{F}_2$ in dynes, between two graphite particles R cm apart have magnitude F_2 given by

$$F_2(R) = -\frac{G}{R^3} + \frac{H}{R^5}, \qquad (14.16)$$

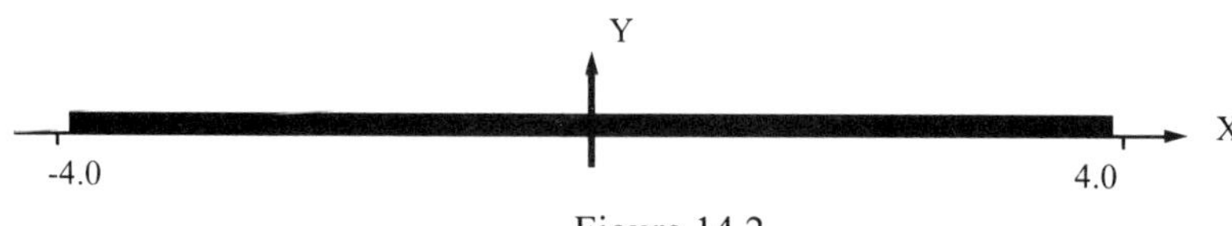

Figure 14.2.

so that

$$\phi_2(R) = -\frac{G}{2R^2} + \frac{H}{4R^4}.\qquad(14.17)$$

Assuming $F_2 = 0$ for two neighbors, (14.16) implies, in analogy with (14.5),

$$-(0.03834)^2 G + H = 0.\qquad(14.18)$$

Assuming zero kinetic energy, the total energy E_2 of the system is approximately

$$E_2 = 3(1003)\left(-\frac{G}{2(0.03834)^2} + \frac{H}{4(0.03834)^4}\right).\qquad(14.19)$$

However, for actual atoms of graphite (Girifalco and Lad (1956), Kelly (1981)) with r in angstroms,

$$\phi(r) = \left(\frac{38591.3}{r^{12}} - \frac{24.3000}{r^6}\right)10^{-12}\text{erg},\qquad(14.20)$$

$$F(r) = \left(\frac{463095.6}{r^{13}} - \frac{145.8}{r^7}\right)10^{-4}\qquad(14.21)$$

where $\vec{F}(r)$ is measured in dynes. In (14.20) and (14.21), $\phi(3.41570) = 0$ and $F(3.83400) = 0$. Then the number N of atoms which fill the slab on a regular triangular mosaic with edge length 3.83400 Å is

$$N = \frac{(7.668)10^8}{3.83400}\frac{(0.13268)10^8}{3.3203483} = (7.992)10^{14}$$

Assuming zero kinetic energy, the total energy E of the system is approximately

$$E = 3(7.992)10^{14}\left(\frac{38591.3}{(3.83400)^{12}} - \frac{24.3000}{(3.83400)^6}\right)10^{-12},\qquad(14.22)$$

so that $E = -9.17149$. Note that the ratio of the numbers in the large parentheses of (14.22) is two. Equating E and E_2 implies

$$-\frac{G}{2(0.03834)^2} + \frac{H}{4(0.03834)^4} = -(3.04802)10^{-3}.\qquad(14.23)$$

The solution of system (14.18) and (14.23) is approximately

$$G = (1.792181)10^{-5} \text{ and } H = (2.634426)10^{-8}$$

Thus

$$F_2 = -\frac{1.792181}{R^3}10^{-5} + \frac{2.634426}{R^5}10^{-8},\qquad(14.24)$$

$$\phi_2 = -\frac{8.960905}{R^2}10^{-6} + \frac{6.586065}{R^4}10^{-9}.\qquad(14.25)$$

Note that $\phi_2(0.0271105) = 0$, so that ϕ_2 can be rewritten as

$$\phi_2 = 4\epsilon_2\left[-\left(\frac{\sigma_2}{R}\right)^2 + \left(\frac{\sigma_2}{R}\right)^4\right]\qquad(14.26)$$

in which $\sigma_2 = 0.271105$ and $\epsilon_2 = (3.048013)10^{-3}$.

Note that since the mass of a carbon atom is approximately $(1.9938)10^{-23}$g, the total atomic mass of the slab, when distributed over the 1003 graphite particles, yields a particle mass M_2 given by

$$M_2 = (1.588679)10^{-11}\text{g}. \tag{14.27}$$

Finally, to determine the force $\vec{F}_3$ between graphite and water particles, we use the empirical bonding law (Hirschfelder, Curtiss and Bird (1965)):

$$\phi_3 = 4\epsilon_3\left[\left(\frac{\sigma_3}{R}\right)^4 - \left(\frac{\sigma_3}{R}\right)^2\right],$$

where $\epsilon_3 = \sqrt{(\epsilon_1, \epsilon_2)} = 0.0113404$ and $\sigma_3 = \frac{1}{2}(\sigma_1 + \sigma_2) = 0.02436945$. Thus

$$\phi_3 = -\frac{2.6938}{R^2}10^{-5} + \frac{1.5998}{R^4}10^{-8}.$$

Hence

$$F_3 = -\frac{5.3877}{R^3}10^{-5} + \frac{6.3992}{R^5}10^{-8}. \tag{14.28}$$

In summary, the water–water interparticle force $\vec{F}_1$, graphite–graphite interparticle force $\vec{F}_2$ and water–graphite interparticle force $\vec{F}_3$ yield, to four significant figures,

$$F_1(R) = -\frac{1.579}{R^3}10^{-4} + \frac{1.477}{R^5}10^{-7}, \tag{14.29}$$

$$F_2(R) = -\frac{1.792}{R^3}10^{-5} + \frac{2.634}{R^5}10^{-8}, \tag{14.30}$$

$$F_3(R) = -\frac{5.388}{R^3}10^{-5} + \frac{6.399}{R^5}10^{-8}, \tag{14.31}$$

while

$$F_1(0.03059) = F_2(0.03834) = F_3(0.03446) = 0. \tag{14.32}$$

The distances $R_1 = 0.03059$, $R_2 = 0.03834$ and $R_3 = 0.03446$ are equilibrium radii.

14.3 Dynamical Equations

In order to derive dynamical equations for particle motions, let us begin with the motion of a water particle P_i as it interacts with other water particles. The motion of P_i, in general, is given by

$$M_1 \frac{d^2 \vec{R}_i}{dt^2} = -980 M_1 + \alpha_1 \sum_j \left(-\frac{1.579}{(R_{ij})^3} 10^{-4} + \frac{1.477}{(R_{ij})^5} 10^{-7} \right) \frac{\vec{R}_{ji}}{R_{ij}} \qquad (14.33)$$

in which the summation is taken over particles P_j which are within a prescribed distance D_1 from P_i, R_{ij} is the distance between P_i and P_j, α_1 is scaling factor which assures that the particle interaction is local relative to gravity, and M_1 is the mass of a water particle given by (14.15). Division by M_1 yields

$$\frac{d^2 \vec{R}_i}{dt^2} = -980 + \alpha_1 \sum_j \left(-\frac{0.61060}{(R_{ij})^3} 10^7 + \frac{0.57115}{(R_{ij})^5} 10^4 \right) \frac{\vec{R}_{ji}}{R_{ij}}. \qquad (14.34)$$

For variety, we now assume, as is common in molecular mechanics, that P_i is acted upon only by particles P_j which are within five equilibrium radii of P_i, so that $D_1 = 5(0.03059) = 0.15295$. By "local relative to gravity" we will now assume the usual 5% experimental error allowance, so that

$$\alpha_1 \left| -\frac{(0.61060)}{(0.15295)^3} 10^7 + \frac{(0.57115)}{(0.15295)^5} 10^4 \right| = 5\%(980),$$

which yields $\alpha_1 = (2.99095)10^{-8}$. Thus (14.34) reduces to

$$\frac{d^2 \vec{R}_i}{dt^2} = -980 + \sum_j \left(-\frac{1.82627}{(R_{ij})^3} 10^{-1} + \frac{1.70828}{(R_{ij})^5} 10^{-4} \right) \frac{\vec{R}_{ji}}{R_{ij}}. \qquad (14.35)$$

Finally, making the changes of variables

$$\bar{R} = 10R, \qquad T = 10t \qquad (14.36)$$

(14.35) reduces to

$$\frac{d^2 \vec{\bar{R}}_i}{dT^2} = -98.0 + \sum_j \left(-\frac{18.2627}{(\bar{R}_{ij})^3} + \frac{1.70828}{(\bar{R}_{ij})^5} \right) \frac{\vec{\bar{R}}_{ji}}{\bar{R}_{ij}}. \qquad (14.37)$$

Using the same line of reasoning, the dynamical equation for the interaction of a graphite particle with other graphite particles is

$$\frac{d^2 \vec{R}_i}{dt^2} = -980 + \sum_j \left(-\frac{0.359575}{(R_{ij})^3} + \frac{5.285275}{(R_{ij})^5} 10^{-4} \right) \frac{\vec{R}_{ji}}{R_{ij}} \qquad (14.38)$$

and the distance of local interaction is $D_2 = 0.1917$. Under transformations (14.36) the equation becomes

$$\frac{d^2 \vec{\bar{R}}_i}{dT^2} = -98.0 + \sum_j \left(-\frac{35.9575}{(\bar{R}_{ij})^3} + \frac{5.285275}{(\bar{R}_{ij})^5} \right) \frac{\vec{\bar{R}}_{ji}}{\bar{R}_{ij}}. \qquad (14.39)$$

The interaction of a water particle P_i with graphite particles P_j is governed by the equation

$$\frac{d^2 \vec{R}_i}{dt^2} = -980 + \sum_j \left(-\frac{0.261085}{(R_{ij})^3} + \frac{3.10075}{(R_{ij})^5} \, 10^{-4} \right) \frac{\vec{R}_{ji}}{R_{ij}} \qquad (14.40)$$

with the local interaction distance $D_3 = 0.1723$. Under transformations (14.36), the equation becomes

$$\frac{d^2 \bar{\vec{R}}_i}{dT^2} = -98.0 + \sum_j \left(-\frac{26.1085}{(\bar{R}_{ij})^3} + \frac{3.10075}{(\bar{R}_{ij})^5} \right) \frac{\bar{\vec{R}}_{ji}}{\bar{R}_{ij}} \qquad (14.41)$$

Note, incidentally, that the first transformation in (14.36) has the effect of multiplying all coordinates in Figs. 14.1 and 14.2 by the factor 10. Hereafter, all discussion is in terms of variables $\bar{R}$ and T. However, the random initial velocities prescribed for particles P_1–P_{823} will be retained.

14.4 Drop and Slab Stabilization

If one uses the leap frog formulas with $\Delta T = 0.00005$ and one allows the water particles to interact in accordance with (14.37), the system exhibits large expansion and contraction modes. The reason is that the initial potential energy is large. To overcome this situation, P_1–P_{823} were allowed to interact in accordance with (14.37), but every 1000 time steps all velocities were damped by the factor 0.9. At the end of 37,000 time steps the damping was removed and the particles were allowed to interact for 9000 more time steps to t_{46000}. The large oscillating modes were no longer present. The resulting stable configuration is shown in Fig. 14.3, where, most importantly, the outermost particles show a lower density than the inner particles, which is characteristic of liquid surface tension. The average diameter in Fig. 14.3 is approximately two-thirds that of Fig. 14.1.

The slab was stabilized in accordance with (14.39), but in the following fashion in order to maintain its solid state. Again we chose $\Delta T = 0.00005$. Whenever the total system kinetic energy exceeded 100, all velocities were damped by the factor 0.25. The system was allowed to run to t_{25000}, at which time the slab had contracted vertically primarily to the relatively stable configuration shown in Fig. 14.4. More extensive time calculations led to crumbling, which was considered to be physically unreasonable. The result in Fig. 14.4 has a height which is approximately two-thirds of that shown in Fig. 14.2.

14.5 Sessile Drop Formation

The drop shown in Fig. 14.3 is now translated vertically upwards 4.5 units so that it sits immediately above the slab shown in Fig. 14.4. This arrangement is shown in Fig. 14.5, with only the central slab particles being plotted. The mode of presentation in this and the next graphs distinguishes between the water and carbon particles so that the interaction can be discerned easily.

The slab particles were allowed no further motion, but P_1–P_{823} were allowed to interact with themselves and with the graphite particles in accordance with (14.37) and (14.39). For the first 200,000 steps of the computer simulation we used $\Delta T = 1(10^{-6})$. Thereafter we used $\Delta T = 2(10^{-6})$. To account for the energy increase due to the effect of gravity, all velocities were damped by 0.9 every 2000 time steps all through the calculations. This was not considered to be significant since our interest was only a relatively steady-state configuration.

The results are summarized in Figs. 14.6–14.9 at the respective times $T = 0.28, 0.60, 0.92,$ and 1.40. The kinetic energy KE is also recorded in each caption since it reflects the motion in the system. The figures show an interlocking of particles below the central fluid mass and a relative steady state at $T = 1.40$. Using linear least square approximations with the four lowest boundary particles on the left and right sides of the system, as shown in Fig. 14.10, we found a left contact angle of 64° and a right contact angle of 68°, the average being 66°. An experimentally determined contact angle measurement reported (Adamson (1976)) is 60°.

Extension of these ideas to three dimensions by Korlie (1966) has yielded a contact angle of 60.01°.

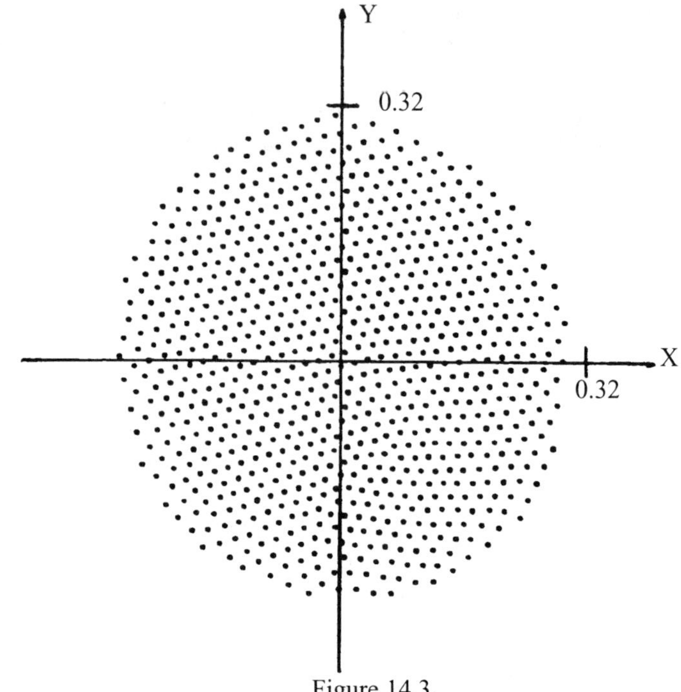

Figure 14.3.

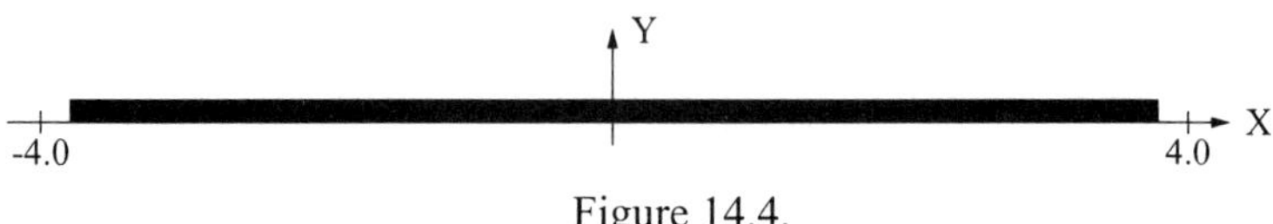

Figure 14.4.

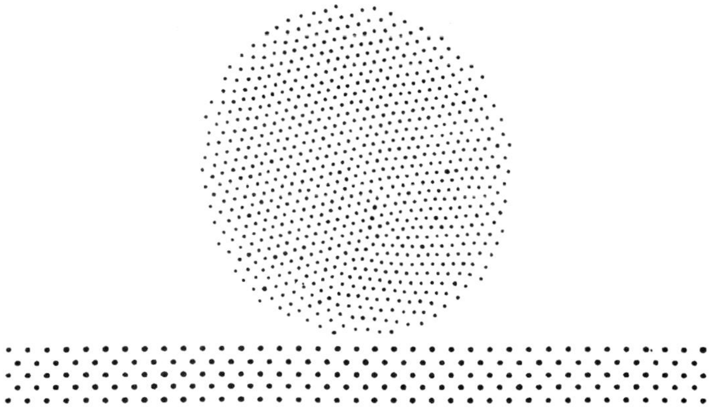

Figure 14.5 — T=0.0.

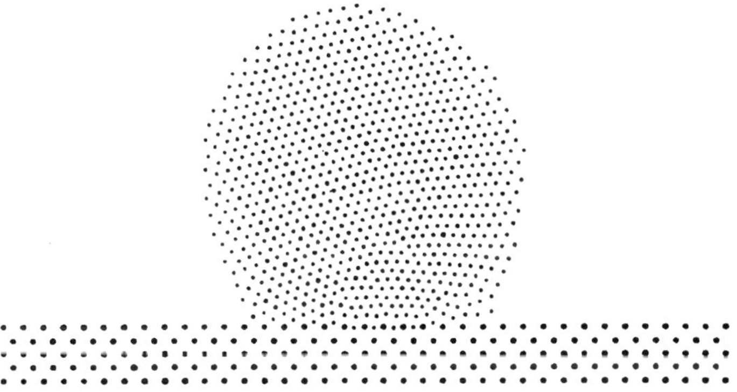

Figure 14.6 — T=0.28, KE=4500.

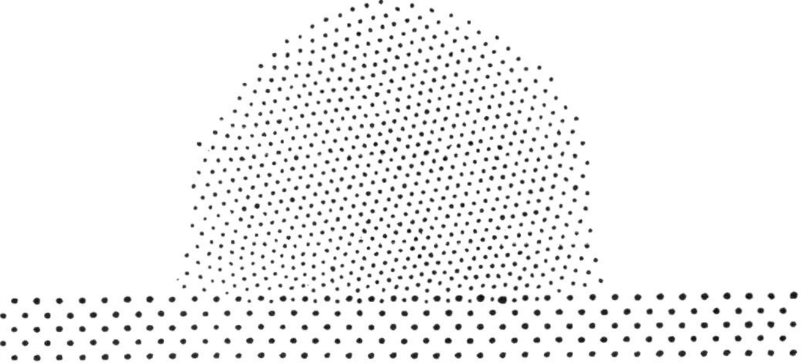

Figure 14.7 — T=0.60, KE=1900.

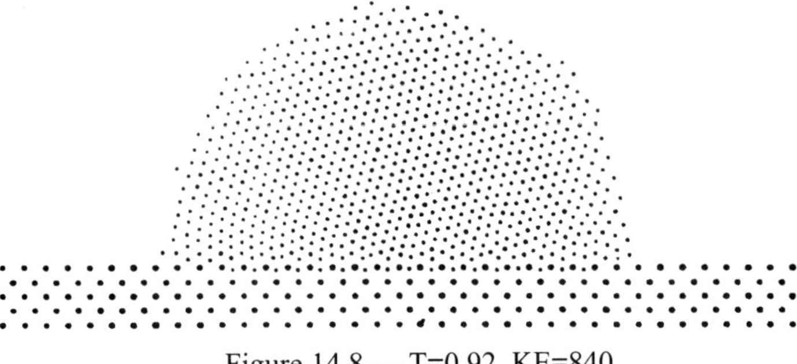

Figure 14.8 — T=0.92, KE=840.

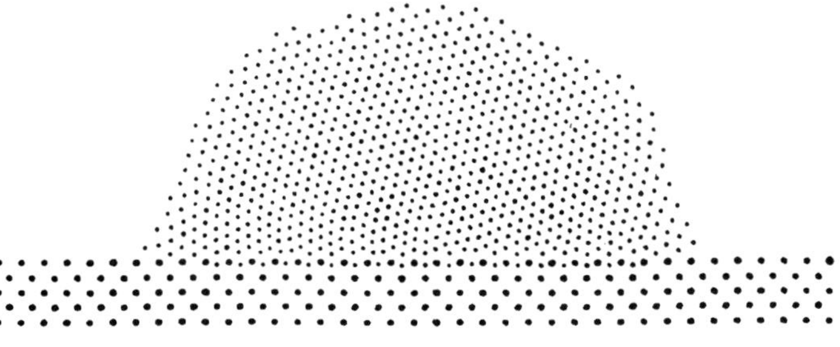

Figure 14.9 — T=1.40, KE=85.

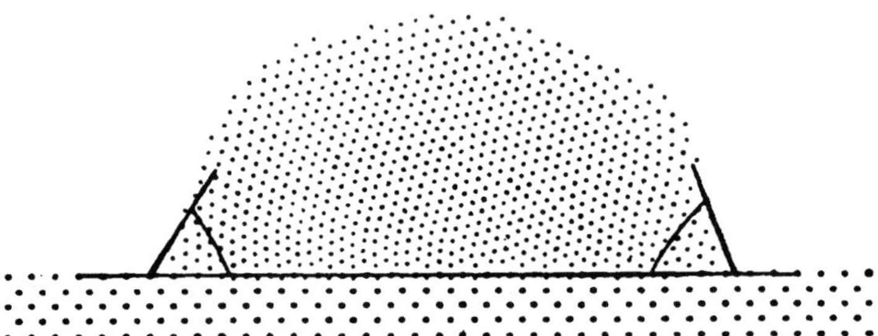

Figure 14.10 — Contact angle fit.

15

Fluid Bubbles and Jiggling Gels

15.1 Introduction

In this chapter, we will develop a particle approach to a two-dimensional study of the motion of carbon dioxide bubbles in water. The discussion for this prototype problem will contain all the concepts and methods which are essential for three-dimensional studies of the motions of fluid drops within fluids.

15.2 Fluid Models

Let us begin by considering first water.

Given two water molecules P_i and P_j, a simplistic classical molecular potential $\phi(r)$ for the pair is the Rowlinson potential (Hirschfelder, Curtiss and Bird (1965)):

$$\phi(r_{ij}) = (2.098)10^{-13}\left[\left(\frac{2.65}{r_{ij}}\right)^{12} - \left(\frac{2.65}{r_{ij}}\right)^{6}\right]\text{erg}, \qquad (15.1)$$

in which the distance r_{ij} between P_i and P_j is measured in angstroms. From (15.1) it follows that the force $\vec{F}_i$ on P_i due to P_j is

$$\vec{F}_i = (2.098)10^{-5}\left[\frac{12(2.65)^{12}}{r_{ij}^{13}} - \frac{6(2.65)^6}{r_{ij}^7}\right]\frac{\vec{r}_{ji}}{r_{ij}}\text{ dynes}. \qquad (15.2)$$

From (15.2) it follows that $F_i(r_{ij}) = \|\vec{F}_i\|$ is given by

$$F_i(r_{ij}) = (2.098)10^{-5}\left[\frac{12(2.65)^{12}}{r_{ij}^{13}} - \frac{6(2.65)^6}{r_{ij}^7}\right], \qquad (15.3)$$

from which it follows that $F_i(\bar{r}) = 0$ implies

$$\bar{r} = 2.975\text{Å},\tag{15.4}$$

which is the equilibrium distance for the force $\vec{F}_i$.

Consider now a two-dimensional, rectangular basin whose base is 23.8 cm and whose height is 23.713 cm. An XY coordinate system is superimposed on the basin as shown in Fig. 15.1. The basin is symmetrical about the Y axis and lies in the upper half plane. On the basin we now construct a regular, triangular grid, as shown in Fig. 15.2. The triangular building block for the grid has edge 1.19 cm and altitude 1.031 cm, as shown in Fig. 15.3. The grid has 24 rows and 492 grid points, which are numbered 1–492 in the usual fashion, left to right on each row and bottom to top.

At each of the constructed grid points, we wish to place a water particle, that is, an aggregate of water molecules. If a point has been numbered i, then the associated particle is denoted by P_i. The mass of each P_i will be determined by distributing equally over the particles the total mass of all the molecules in the basin. Use of (15.4) as the edge of a regular, triangular grid of water molecules implies that the number N of water molecules in the basin is, approximately,

$$N = \frac{(23.8)(23.713)}{(2.975)10^{-8}(2.576)10^{-8}} = (7.364)10^{17}.\tag{15.5}$$

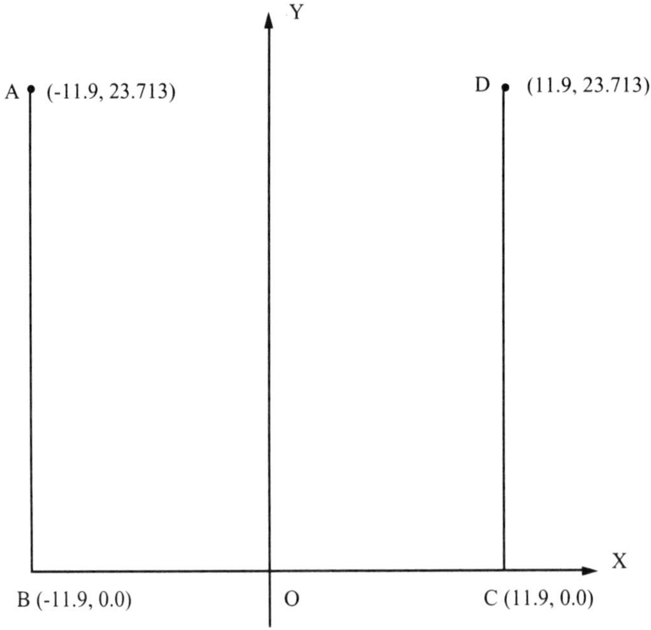

Figure 15.1.

Figure 15.2 — Triangular basin grid.

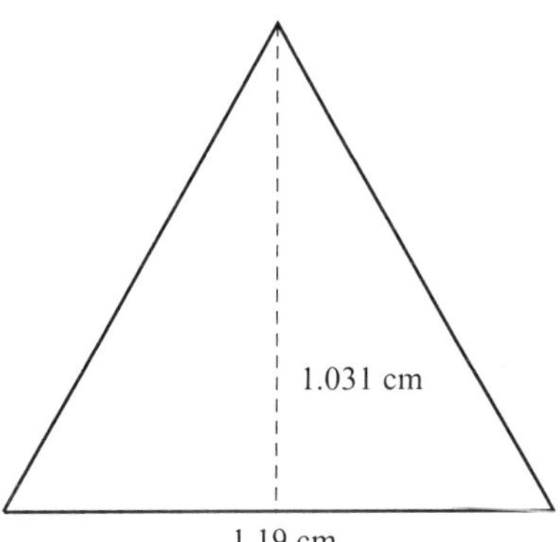

Figure 15.3 — Regular triangular building block.

Now, the mass of a single H_2O molecule is $(30.103)10^{-24}$ gr, so that the total H_2O molecular mass is $(2.217)10^{-5}$ gr. Distributing this over the 492 particles implies that the mass M_1 of an H_2O particle is

$$M_1 = (4.506)10^{-8} \text{gr}. \tag{15.6}$$

Next we wish to develop a formula for the force between two particles P_i and P_j. For variety, we assume in this chapter that the force has a magnitude

$$F = \frac{G}{R_{ij}} + \frac{H}{R_{ij}^3} \text{ dynes}, \tag{15.7}$$

in which the distance R_{ij} between P_i and P_j is measured in cm. From (15.7) then,

$$\phi(R_{ij}) = -G \log R_{ij} + \frac{H}{2R_{ij}^2}. \tag{15.8}$$

Assuming that the basic edge length 1.19 cm of the triangular grid in Fig. 15.2 is the equilibrium distance for the force magnitude F in (15.7) implies

$$\frac{G}{1.19} + \frac{H}{(1.19)^3} = 0. \tag{15.9}$$

A second equation for G and H is determined by computing potential energies of the particle and molecular systems as follows. Assuming that all velocities are zero, the potential energy E of the particle system is, approximately,

$$E = 3 \sum_{i=1}^{492} \left(-G \log 1.19 + \frac{H}{2(1.19)^2} \right), \tag{15.10}$$

or,

$$E = 3(492)(-0.17395G + 0.35308H), \tag{15.11}$$

while that for the molecular system is, approximately,

$$E = 3 \sum_{i=1}^{(7.364)10^{17}} \left\{ (2.098)10^{-13} \left[\left(\frac{2.65}{2.975} \right)^{12} - \left(\frac{2.65}{2.975} \right)^{6} \right] \right\}, \tag{15.12}$$

so that

$$E = -(1.159)10^5 \text{erg}. \tag{15.13}$$

Thus, (15.11) and (15.13) imply

$$-0.4927G + H = -222.4. \tag{15.14}$$

Finally, the solution of (15.9) and (15.14) is

$$G = 116.51, \qquad H = -164.99 \tag{15.15}$$

so that (15.7) becomes

$$F = \frac{116.51}{R_{ij}} - \frac{164.99}{R_{ij}^3}. \tag{15.16}$$

Next, let us develop dynamical equations for H_2O particles. From (15.6) and (15.16), let the motion of P_i be determined by the dynamical equation

$$(4.506)10^{-8}\frac{d^2\vec{R}_i}{dt^2} = -980(4.506)10^{-8}\vec{\delta}$$

$$+ \alpha \sum_{\substack{j=1 \\ j\neq i}}^{492} \left[\frac{116.51}{R_{ij}} - \frac{164.99}{R_{ij}^3}\right]\frac{\vec{R}_{ji}}{R_{ij}} \tag{15.17}$$

in which $\vec{\delta} = (0, 1)$ and α is a normalization constant which is determined as follows. From (15.17)

$$\frac{d^2\vec{R}_i}{dt^2} = -980\vec{\delta} + \alpha \sum_{\substack{j=1 \\ j\neq i}}^{492} \left(\frac{25.86}{R_{ij}} - \frac{36.26}{R_{ij}^3}\right)10^8\frac{\vec{R}_{ji}}{R_{ij}}. \tag{15.18}$$

The normalization constant α is now chosen so that each particle P_i in the very top row in Fig. 15.2 is supported very strongly by the local interaction with any particle which lies 1.0305 units directly below it (see Fig. 15.3), which we interpret to imply

$$\frac{1}{2}\alpha\left(\frac{25.86}{1.0305} - \frac{36.62}{1.0305^3}\right)10^8 = 980. \tag{15.19}$$

Thus, one finds

$$\alpha = (-234.19)10^{-8}. \tag{15.20}$$

Hence, (15.18) reduces to

$$\frac{d^2\vec{R}_i}{dt^2} = -980\vec{\delta} + \sum_{\substack{j=1 \\ j\neq i}}^{492} \left(-\frac{6058}{R_{ij}} + \frac{8576}{R_{ij}^3}\right)\frac{\vec{R}_{ji}}{R_{ij}}. \tag{15.21}$$

For actual computation, we will make the convenient change of variables

$$T = 10t, \tag{15.22}$$

so that (15.20) becomes finally

$$\frac{d^2\vec{R}_i}{dT^2} = -9.8\vec{\delta} + \sum_{\substack{j=1 \\ j\neq i}}^{492} \left[-\frac{60.56}{R_{ij}} + \frac{85.76}{R_{ij}^3}\right]\frac{\vec{R}_{ji}}{R_{ij}} \tag{15.23}$$

15.3 Basin Stabilization

We next let a basin of H_2O particles find its own equilibrium configuration dynamically as follows. Consider then a basin of H_2O particles at the grid points of the triangular grid shown in Fig. 15.2. To avoid symmetry, a velocity of either ± 0.0000001 is assigned in the X direction, at random, to each particle. From these initial data, the motion of the system is generated as follows. The dynamical equation of each P_i is taken to be

$$\frac{d^2 \vec{R}_i}{dT^2} = -9.8\vec{\delta} + \sum_{\substack{j=1 \\ j \neq i}}^{492} \left[-\frac{A}{R_{ij}} + \frac{B}{R_{ij}^3} \right] \frac{\vec{R}_{ji}}{R_{ij}} , \qquad (15.24)$$

in which $R_{ij} > 1.2$ implies $A = B = 0$, while $R_{ij} \leq 1.2$ implies $A = 60.56$, $B = 85.76$. In this fashion, interparticle force is kept strictly local. The resulting 492-body problem is solved numerically by the leap frog formulas with $\Delta T = 0.0002$. Every 500 time steps, each velocity is damped by the factor 0.1. Particle reflection due to wall collision is done symmetrically with velocity damped by the factor 0.1. In the usual notation $T_k = k\Delta T$, $k = 0, 1, 2, \ldots$, the evolution of the basin through T_{140000} is shown in Figs. 15.4–15.6.

Calculations were halted at T_{140000} because we desired to have a fluid with nontrivial internal and surface motions.

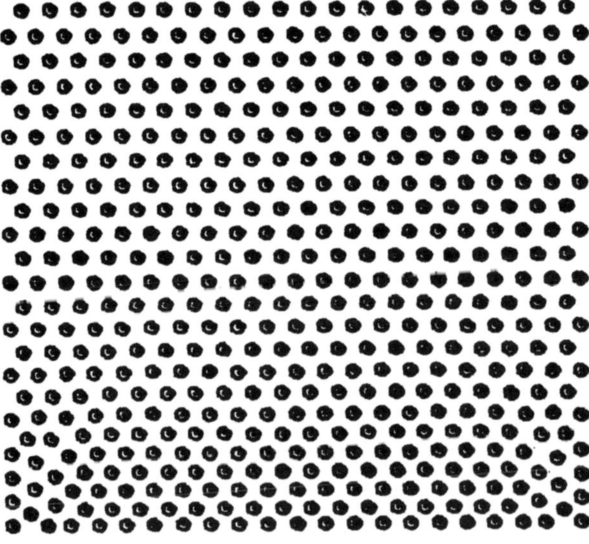

Figure 15.4 — H_2O basin at T_{20000}.

Figure 15.5 — H_2O basin at T_{80000}.

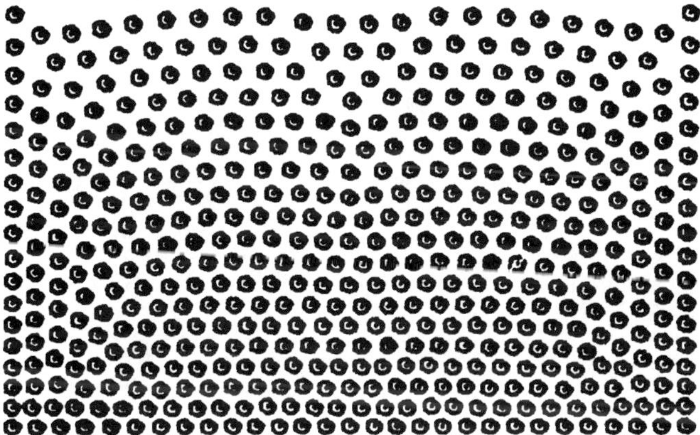

Figure 15.6 — H_2O basin at T_{140000} (enlarged)

15.4 Motion of CO_2 Bubbles

To simulate the motion of CO_2 bubbles in H_2O, we must first repeat for a CO_2 gas the considerations in Sect. 15.2. Since CO_2 gas in three dimensions at 0°C has a density approximately (Sears and Zemansky (1957)) 1/500 that of H_2O in two dimensions the density will be approximately $500^{-2/3}$, or, approximately 1/63 that of H_2O. Thus, into the basin shown in Fig. 15.1 we place $(1.169)10^{16}$ molecules and 492/63, or, for convenience, seven CO_2 particles. The particles are arranged as shown in Fig. 15.7.

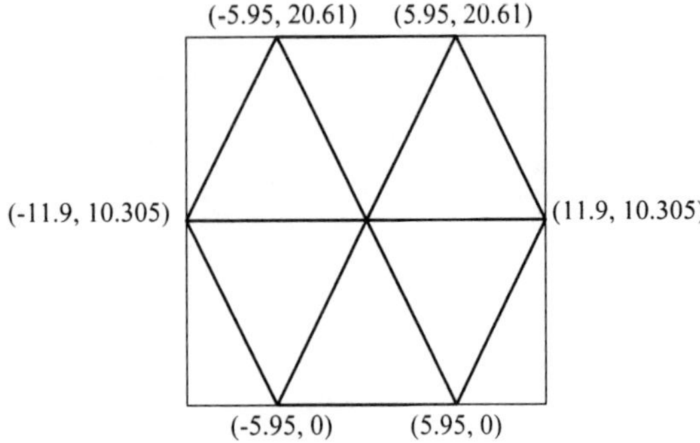

Figure 15.7 — CO_2 particles

For the molecular arrangement, note that a potential for CO_2 is (Hirschfelder, Curtiss and Bird (1965)):

$$\phi(r_{ij}) = (1.132051)10^{-13}\left[\left(\frac{4.07}{r_{ij}}\right)^{12} - \left(\frac{4.07}{r_{ij}}\right)^{6}\right], \tag{15.25}$$

so that

$$F(r_{ij}) = (1.132051)10^{-5}\left[\frac{12(4.07)^{12}}{r_{ij}^{13}} - \frac{6(4.07)^{6}}{r_{ij}^{7}}\right], \tag{15.26}$$

which yields an equilibrium distance of 4.57Å. One then has an approximate total molecular potential energy of

$$E = 3\sum_{1}^{(1.17)10^{16}}\left\{(1.132051)10^{-13}\left[\left(\frac{4.07}{4.57}\right)^{12} - \left(\frac{4.07}{4.57}\right)^{6}\right]\right\} \tag{15.27}$$

or

$$E = -993.4 \text{ erg}. \tag{15.28}$$

For CO_2 particles, we assume

$$F = \frac{A}{R_{ij}} + \frac{B}{R_{ij}^{3}}, \tag{15.29}$$

with R_{ij} measured in cm. Then,

$$\phi = -A\log R_{ij} + \frac{B}{2R_{ij}^{2}}. \tag{15.30}$$

Assuming that $R = 11.9$ cm is the CO_2 particle equilibrium distance implies

$$\frac{A}{11.9} + \frac{B}{11.9^{3}} = 0. \tag{15.31}$$

The total potential energy of the particle system is, approximately,

$$E = 12\left(-A \log 11.9 + \frac{B}{2(11.9)^2}\right),$$

so that

$$12\left(-A \log 11.9 + \frac{B}{2(11.9)^2}\right) = -993.4. \tag{15.32}$$

The solution of (15.31) and (15.33) is $A = 27.800$, $B = -3936.9$.

Finally, since the mass of a CO_2 molecule is $(7.3585)10^{-23}$gr, the total molecular mass M is

$$M = (7.3585)10^{-23}(1.169)10^{16} = (8.602)10^{-7}\text{gr} \tag{15.33}$$

and the mass M_3 of a CO_2 particle is

$$M_3 = M/7 = (1.2289)10^{-7}\text{gr}. \tag{15.34}$$

Thus, the equation of a CO_2 particle is

$$M_3 \frac{d^2 \vec{R}_i}{dt^2} = -980\vec{\delta} M_3 + \alpha \sum \left[\frac{27.800}{R_{ij}} - \frac{3936.9}{R_{ij}^3}\right]\frac{\vec{R}_{ji}}{R_{ij}}. \tag{15.35}$$

From (15.20) and (15.34), then,

$$\frac{d^2 \vec{R}_i}{dt^2} = -980\vec{\delta} + \frac{(-234.2)10^{-8}}{(1.2289)10^{-7}} \sum \left[\frac{27.800}{R_{ij}} - \frac{3936.9}{R_{ij}^3}\right]\frac{\vec{R}_{ji}}{R_{ij}} \tag{15.36}$$

or,

$$\frac{d^2 \vec{R}_i}{dt^2} = -980\vec{\delta} + \sum \left[-\frac{529.78}{R_{ij}} + \frac{75025}{R_{ij}^3}\right]\frac{\vec{R}_{ji}}{R_{ij}}. \tag{15.37}$$

Hence, by (15.22) we find

$$\frac{d^2 \vec{R}_i}{dT^2} = -9.8\vec{\delta} + \sum \left[-\frac{5.298}{R_{ij}} + \frac{750.25}{R_{ij}^3}\right]\frac{\vec{R}_{ji}}{R_{ij}}. \tag{15.38}$$

We need next equations of motion for CO_2–H_2O interaction. For H_2O–H_2O and CO_2-CO_2 particle interactions, the equations are (15.23) and (15.38), respectively. For CO_2–H_2O particle interaction, we use an ultrasimplistic law of empirical bonding in which the local interaction constants are averaged. However, we will also impose a local interaction distance D to force local interaction only. Our choice is $D = 1.2$. Thus the following dynamical approach will be used. Let P_i and P_j be any two particles in the basin shown in Fig. 15.6. The motion of P_i is determined by the dynamical equation

$$\frac{d^2 \vec{R}_i}{dT^2} = -9.8\vec{\delta} + \sum_{\substack{j=1 \\ j\neq i}}^{492} \left[-\frac{A}{R_{ij}} + \frac{B}{R_{ij}^3}\right]\frac{\vec{R}_{ji}}{R_{ij}} \tag{15.39}$$

If $R_{ij} > 1.2$, then $A = B = 0$. If $R_{ij} \leq 1.2$, then A and B are determined as follows. If P_i, P_j are both H_2O particles, then $A = 60.56$, $B = 85.76$. If P_i, P_j are both CO_2 particles, then $A = 5.30$, $B = 750.25$. In all other cases, $A = 32.93 = \frac{1}{2}(60.56 + 5.30)$, $B = 418.01 = \frac{1}{2}(85.76 + 750.25)$.

As a first example, consider the H_2O basin shown in Fig. 15.6. The particles P_{108}, P_{158}, P_{213}, P_{263}, P_{294}, P_{298}, and P_{362} are now assumed to be CO_2 particles. No changes in positions or velocities are made. The initial configuration is shown in Fig. 15.8. The system (15.39) was solved numerically with $\Delta T = 0.0002$ by the leap frog formulas through T_{640000}. The natural, rapid bubble emergence from the basin is shown in Figs. 15.9–15.14 at the indicated times.

As a second example, the first example was repeated in each detail with the single exception that the basin used was a D_2O basin. The emergence of the bubbles from the basin was, approximately, 0.7 times faster than from the H_2O basin.

Consider next setting the seven CO_2 particles in the H_2O basin in the positions P_{216}, P_{217}, P_{236}, P_{237}, P_{238}, P_{257}, P_{258} as shown in Fig. 15.15. The effect is to have created a large compressed gas bubble. One must now expect the generation of a compression wave. With $\Delta T = 0.00002$, the resulting motion is shown in Figs. 15.16–15.20 at the indicated times. Fig. 15.16 shows the immediate compression wave effect directly above the bubble at the basin surface. The figures also show the disintegration of the bubble as it rises. Fig. 15.21 shows at T_{160000} only those H_2O particles which were originally below the bubble and their formation into a wake below the CO_2 as it rises. Fig. 15.22 shows at this time how the particles originally at the top of the basin have moved downward toward the area vacated by particles in the wake. A large rotational H_2O motion is evident at this time.

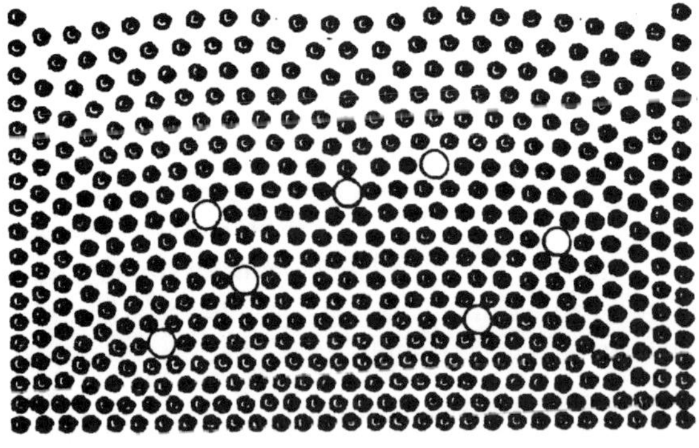

Figure 15.8 — Initial $CO_2 - H_2O$ configuration.

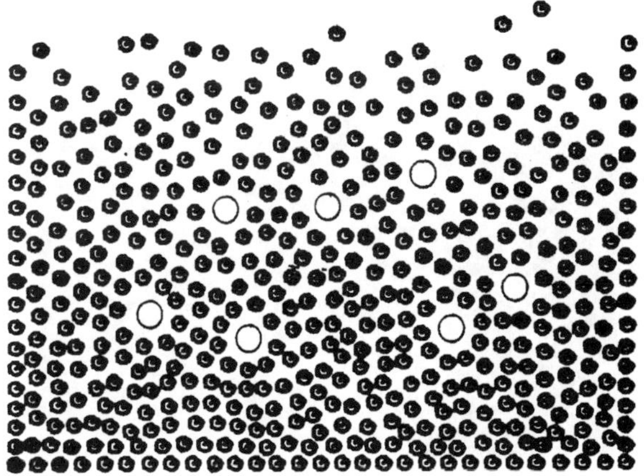

Figure 15.9 — $T = T_{20000}$.

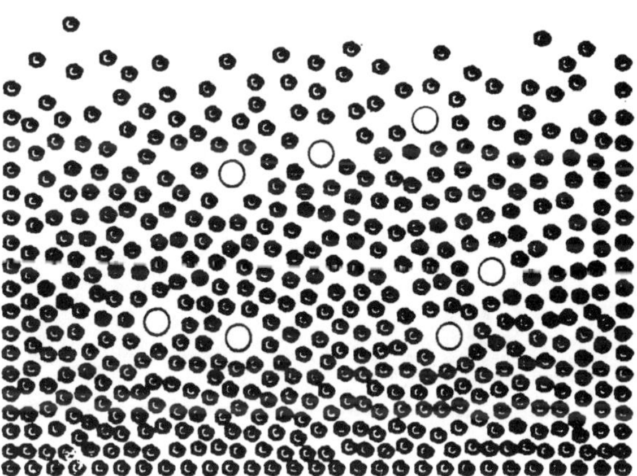

Figure 15.10 — $T = T_{40000}$.

15.5 Jiggling Gels

Solids have been classified in a variety of ways, as being, for example, strong, weak, hard, brittle, soft, ductile, and elastic (see, e.g., Burridge (1978), Kelly and Macmillan (1986)). In this final section we will discuss, in a quasiquantitative fashion, solids like gels, which shake when a moderate external force is applied. We will consider only hydrophilic H_2O-based gels so that the discussion in Sect. 15.2 will be of special value.

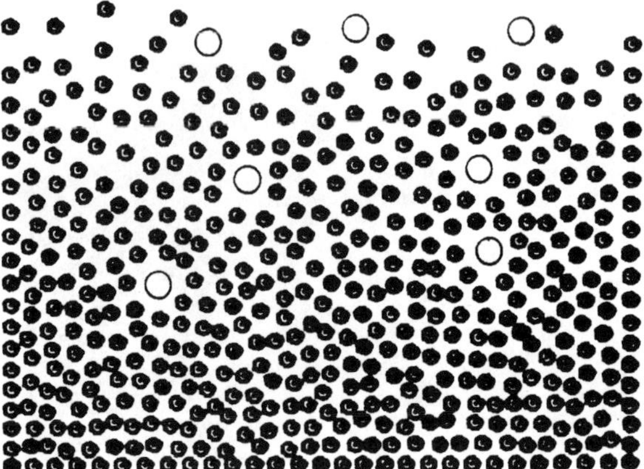

Figure 15.11 — $T = T_{80000}$.

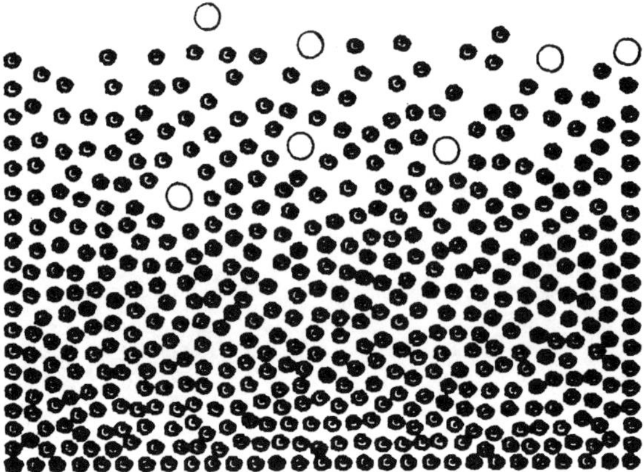

Figure 15.12 — $T = T_{120000}$.

Consider (15.17), that is

$$(4.506)10^{-8}\frac{d^2\vec{R}_i}{dT^2} = -980(4.506)10^{-8}\vec{\delta}$$

$$+ \alpha \sum_{\substack{j=1 \\ j\neq i}}^{492} \left[\frac{116.51}{R_{ij}} - \frac{164.99}{R_{ij}^3} \right] \frac{\vec{R}_{ji}}{R_{ij}} \qquad (15.40)$$

Next, let us not consider all the particles shown in Fig. 15.2 but only those in the regular "conical" subregion called a cone, shown in Fig. 15.23. This conical subregion contains 231 particles which are now renumbered in the usual fashion,

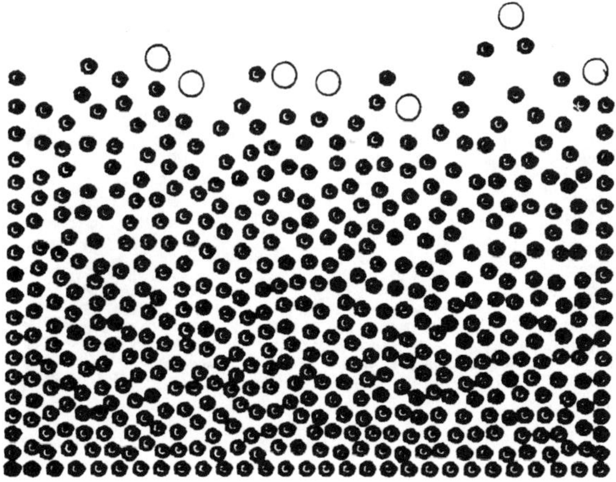

Figure 15.13 — $T = T_{180000}$.

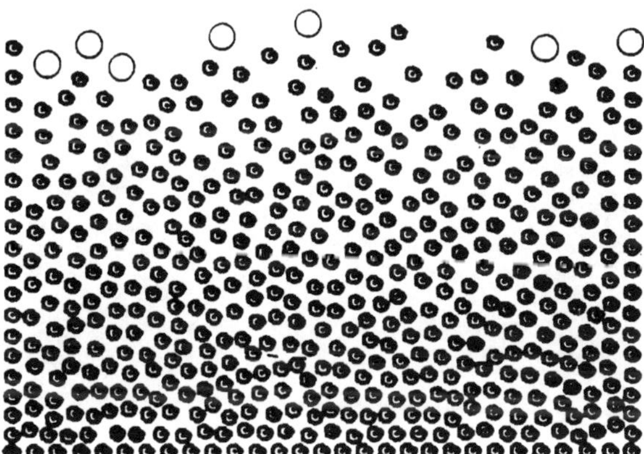

Figure 15.14 — $T = T_{240000}$.

left to right and bottom to top. Thus, (15.40) now becomes

$$(4.506)10^{-8}\frac{d^2\vec{R}_i}{dT^2} = -980(4.506)10^{-8}\vec{\delta} + \alpha \sum_{\substack{j=1\\j\neq i}}^{231}\left[\frac{116.51}{R_{ij}} - \frac{164.99}{R_{ij}^3}\right]\frac{\vec{R}_{ji}}{R_{ij}}$$

which can be simplified further to

$$\frac{d^2\vec{R}_i}{dt^2} = -980\vec{\delta} + \alpha \sum_{\substack{j=1\\j\neq i}}^{231}\left\{\left[\frac{25.86}{R_{ij}} - \frac{36.62}{R_{ij}^3}\right]10^8\right\}\frac{\vec{R}_{ji}}{R_{ij}} \tag{15.41}$$

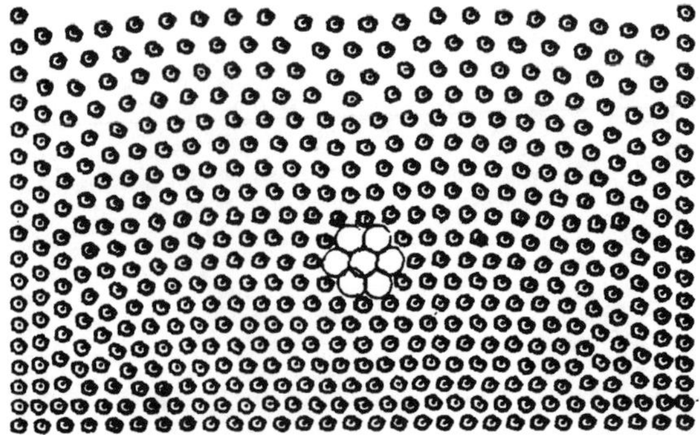

Figure 15.15 — Initial $CO_2 - H_2O$ configuration.

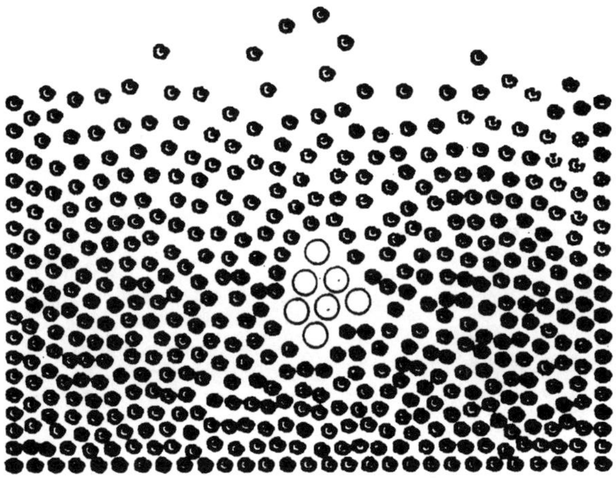

Figure 15.16 — $T = T_{20000}$.

For computer implementation, we make the convenient change of variables

$$T = 10t$$

so that (15.41) becomes

$$\frac{d^2 \vec{R}_i}{dT^2} = -9.8 + \alpha \sum_{\substack{j=1 \\ j \neq i}}^{231} \left\{ \left[\frac{0.2586}{R_{ij}} - \frac{0.3662}{R_{ij}^3} \right] (10)^8 \right\} \frac{\vec{R}_{ji}}{R_{ij}} \qquad (15.42)$$

We now turn to water-based gels, that is, to hydrophilic solids combined with water to form a highly flexible solid. Since intermolecular forces are not readily available for classes of gels, we will approximate such solids by choosing α in

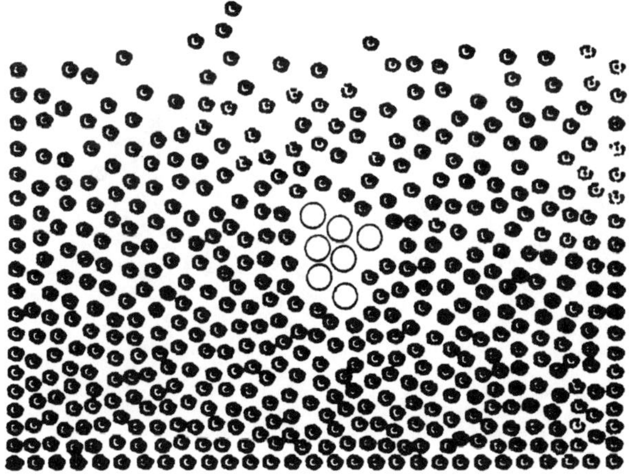

Figure 15.17 — $T = T_{60000}$.

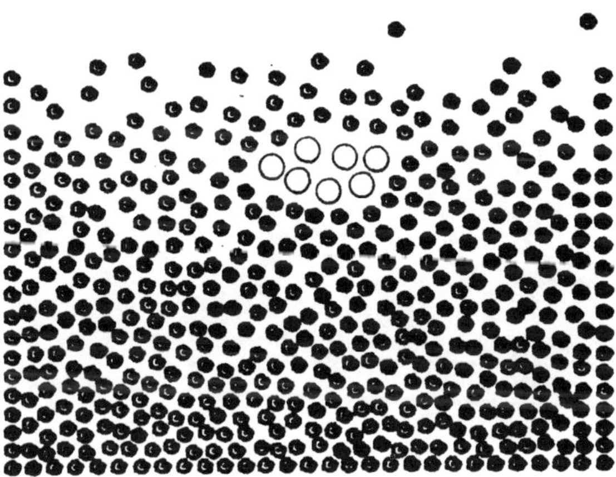

Figure 15.18 — $T = T_{160000}$.

(15.42) so that the resulting solids exhibit various degrees of flexibility. For this purpose, we consider the following five cases:

Gel A: $\alpha = -3(10^{-4})$

Gel B: $\alpha = -2(10^{-4})$

Gel C: $\alpha = -1(10^{-4})$

Gel D: $\alpha = -0.5(10^{-4})$

Gel E: $\alpha = -0.25(10^{-4})$

The dynamical equations for the gels are (15.42) with the corresponding choice for α. The initial structure first presented in Fig. 15.23 is now given the particle representation shown in Fig. 15.24. All of its position coordinates are known.

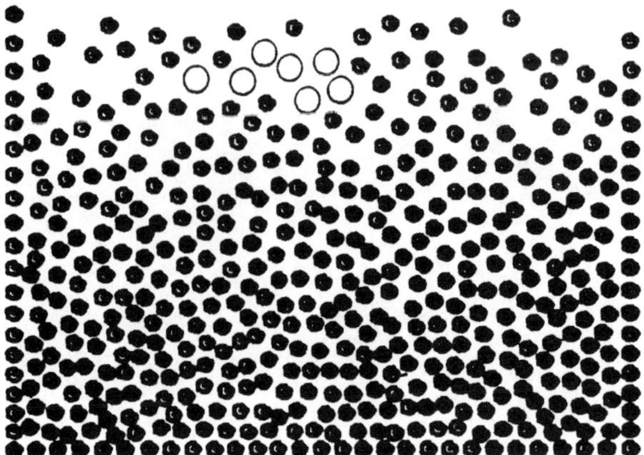

Figure 15.19 — $T = T_{240000}$.

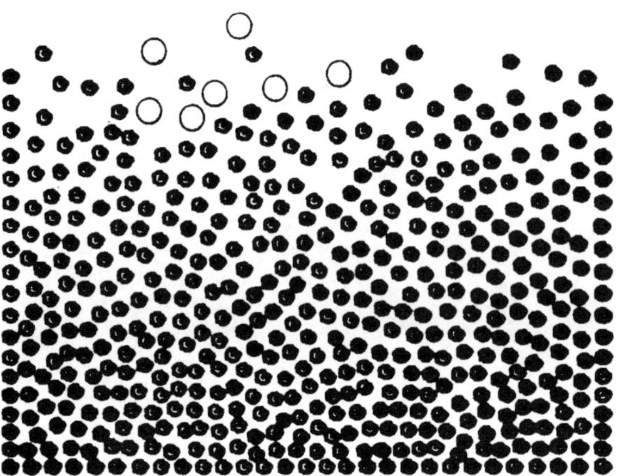

Figure 15.20 — $T = T_{340000}$.

We then assign to each particle, at random, a small, symmetry upsetting, velocity component in the X direction whose values are 0 or $\pm 10^{-10}$, so that all initial data are available.

The system (15.42) for the 231-body problem is now solved numerically by the leap frog formulas with $\Delta T = 0.0001$. In this fashion, the system will find its own stable equilibrium form.

Two important aspects of the computation are as follows. First, if two particles are separated by more than 1.6 cm, then the force between them is taken to be zero. Second, the base particles are allowed to be supported by the X axis. Thus, P_1–P_{21} are reflected symmetrically with respect to the X axis. This is done in such

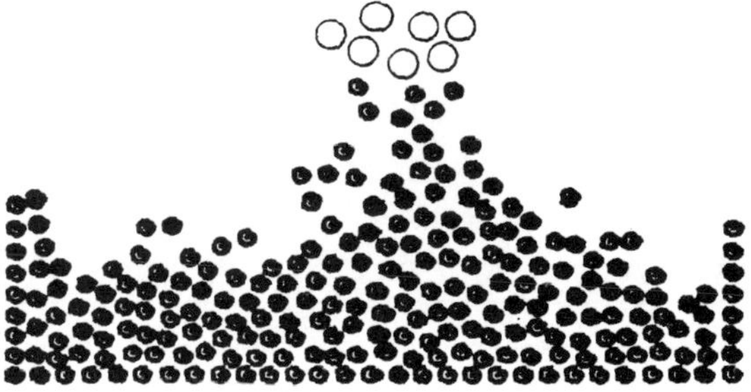

Figure 15.21 — Wake flow at T_{160000}.

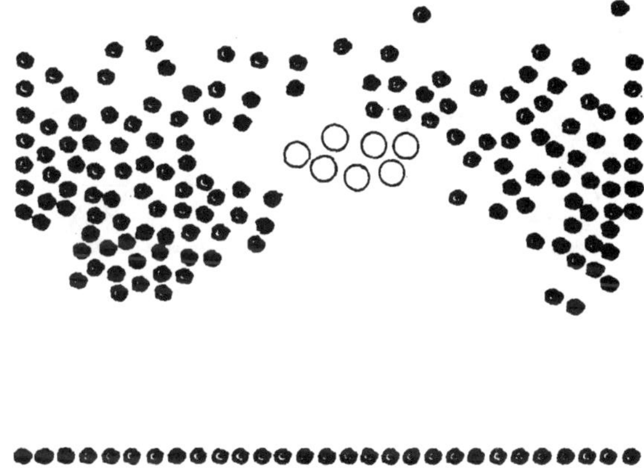

Figure 15.22 — Vertical flow of uppermost particles at T_{160000}.

a fashion that their velocity components become $v_y \to -0.1v_y$, $v_x \to 0$. Thus, strong friction is imposed on the base.

The results after 12 million time steps for gels A–C are shown in Fig. 15.25 A–C, while those after 20 million time steps for gels D and E are shown in Fig. 15.25 D–E. The total kinetic energy for the stabilized forms in Fig. 15.25 A–C is less than unity, while for D and E it is less than ten. Further calculation with gels D and E did not reduce the kinetic energy of either significantly.

Though the results in Fig. 15.25 look almost the same, there are differences. The resulting height h and base width d for each is

Gel A: $h = 20.524$ cm, $d = 23.858$ cm
Gel B: $h = 20.481$ cm, $d = 23.888$ cm
Gel C: $h = 20.349$ cm, $d = 23.982$ cm
Gel D: $h = 20.109$ cm, $d = 24.180$ cm
Gel E: $h = 19.618$ cm, $d = 24.639$ cm

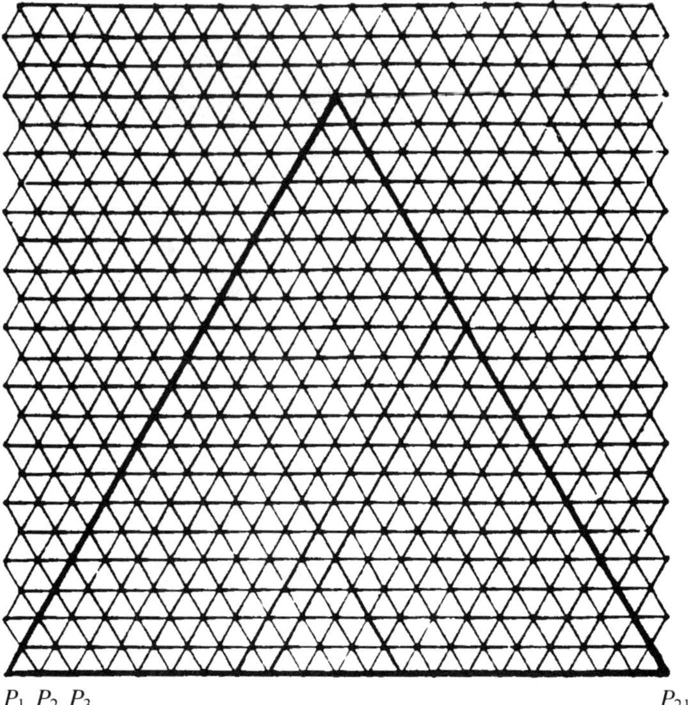

$P_1\ P_2\ P_3$ P_{21}

Figure 15.23.

Figure 15.24 — Initial cone.

The fact that h decreases with $|\alpha|$ while d increases with $|\alpha|$ is to be expected, since the effect of gravity increases with the decrease in $|\alpha|$.

Note finally that the choice $\alpha = -(0.1)10^{-4}$ is not adequate to maintain solid structure and results in all the particles falling to the X axis.

As shown in Fig. 15.26 the particles P_1, P_{90}, P_{140}, and P_{168} have been removed from the stabilized gels A and E, creating a rounded corner at the left end of the base, a wedge on the right edge, and two holes in the core. The dynamical implementation described above is now continued as a 227-body problem. Fig. 15.27 shows the results at T_{15000}. No changes result after this time.

Comparison of Figs. 15.25a and 15.27 reveals that for gel A, which has the largest $|\alpha|$ of all the gels, there is no distortion due to the loss of particles. For gel E, which has the smallest $|\alpha|$, there is a small compressive result where particles have been removed. In both cases, however, the solid structure is maintained. For the five gels A–E, the amount of compression increases as $|\alpha|$ decreases, while continuing to maintain solid structure.

Jiggling is the large elastic motion which a gel exhibits when acted upon by an external force. To demonstrate jiggling, let us consider gel C. Let us first tip its base so that the base forms an angle of $15°$ with the X axis, as shown in Fig. 15.28 at $T = T_0$. The gel is allowed to fall freely from this initial position. When particles collide with the X axis, the reflection rule used is $x_k \rightarrow x_k$, $y_k \rightarrow -y_k$, $v_{k,x} \rightarrow 0$, $v_{k,y} \rightarrow 0$. Fig. 15.28 then shows the motion of the system every 3000 time steps through T_{39000}. The jiggling motion is evident throughout. That at the apex is particularly vivid from T_{9000} to T_{15000}. By T_{39000} the extent of the jiggling is beginning to dissipate.

Gels A and B exhibit similar behavior to that of C but with less extensive and more rigid motions. Gels D and E first show crumbling due to the weight on P_1 and the increased effect of gravity due to the smaller values of $|\alpha|$. Thereafter, however, they do jiggle.

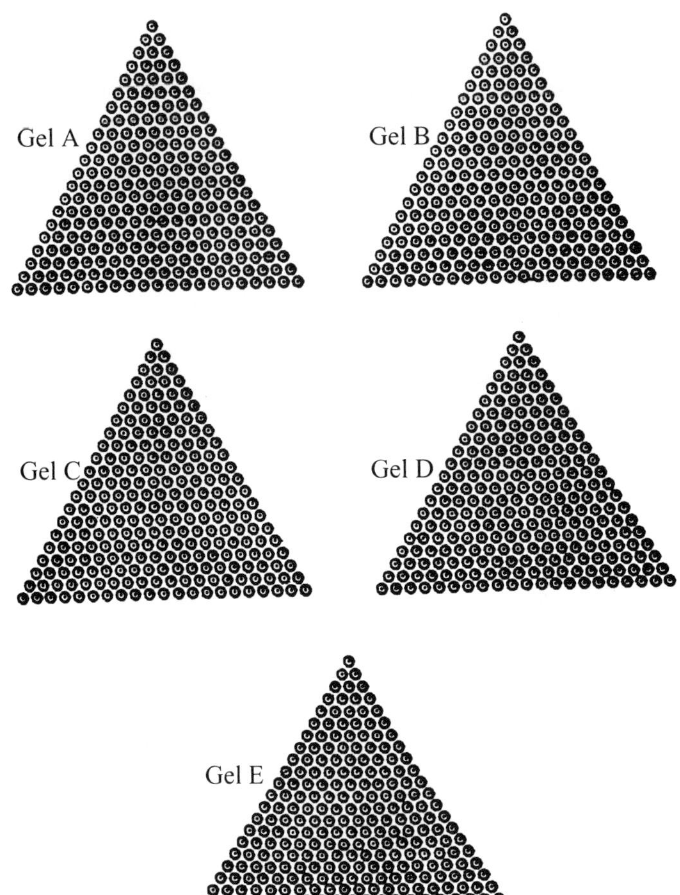

Figure 15.25 — Stabilized forms of gels A–E.

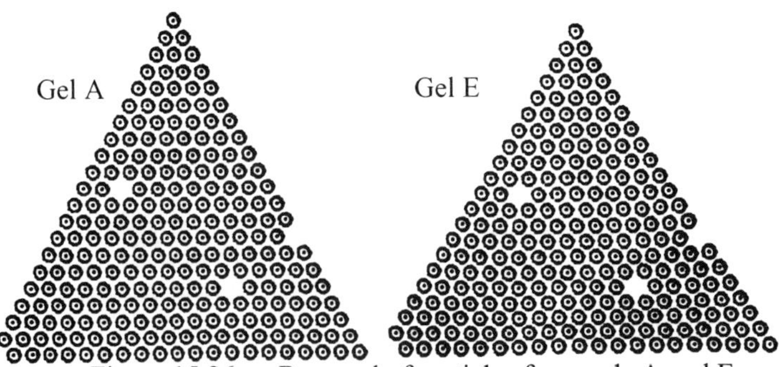

Figure 15.26 — Removal of particles from gels A and E.

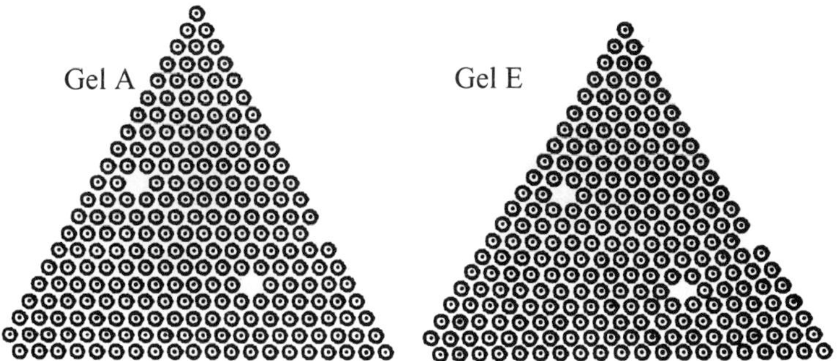

Figure 15.27 — Results of particle removal from gels A and E.

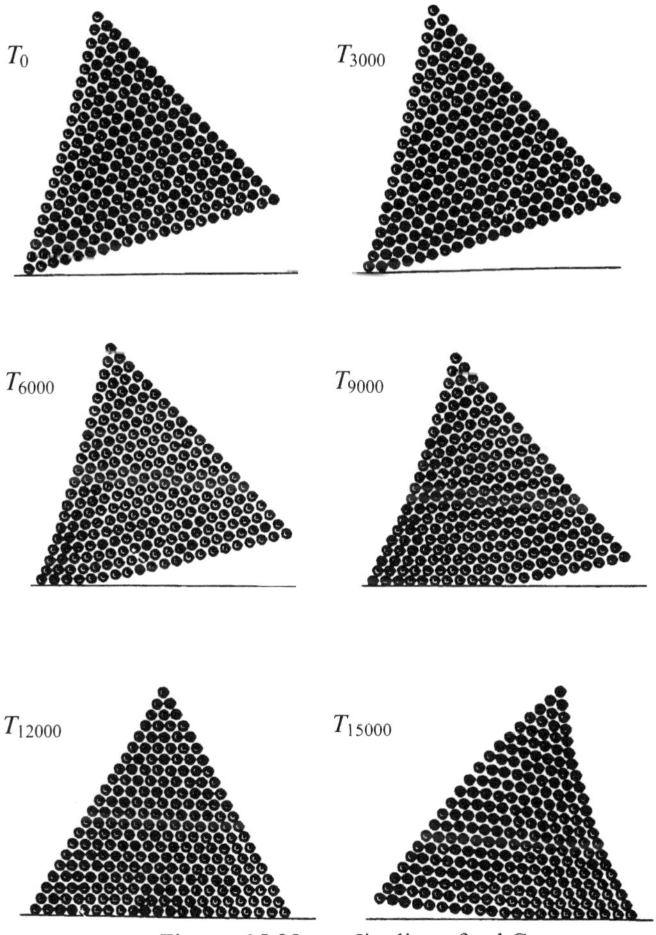

Figures 15.28a — Jiggling of gel C.

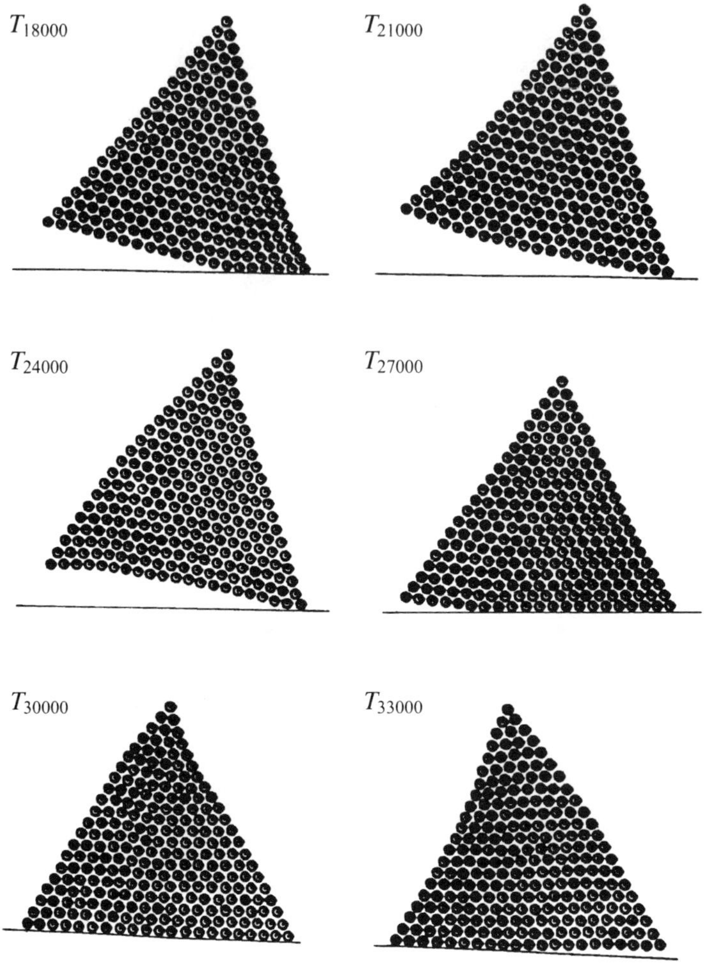

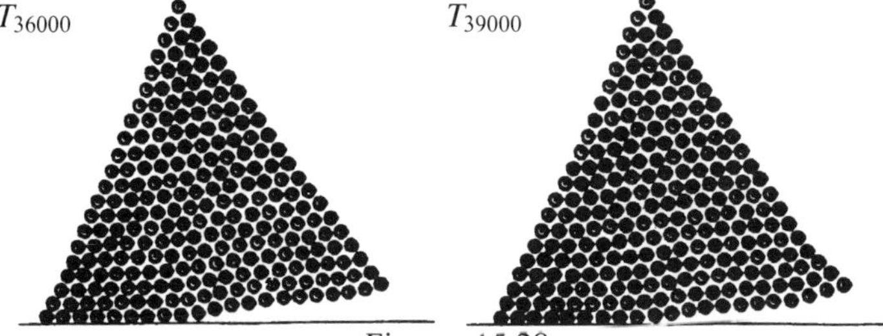

Figures 15.28b.

Figures 15.28c.

16

Melting Points

16.1 Introduction

The melting point of a solid characterizes, in a fundamental way, transition between solid and fluid states. It is usually defined in terms of the average kinetic energy of a *large* ensemble of atoms or molecules (Cotterill, Kristensen and Jensen (1974), Hirschfelder, Curtiss and Bird (1965)). In this chapter, we will develop an approach for determining the melting point of a homogeneous atomic or molecular solid via the four-body problem. As a consequence of our approach, a new formula in terms of Planck's constant, rather than Boltzmann's constant, results.

16.2 Formula Development

Consider first four identical atoms P_1, P_2, P_3, P_4, each of mass m. Let $\phi(r)$ be a related classical interatomic potential and let $\vec{F}$ be the interatomic force defined by ϕ. Let $\vec{F}$ be zero when r equals r^*, the equilibrium distance. Though r and r^* will be given in angstroms, since this is customary, all other quantities will be given in cgs units.

Next set P_i, $i = 1, 2, 3, 4$, to be the vertices of a regular tetrahedron of edge length r^*, at the respective points (x_i, y_i, z_i), as shown in Fig. 16.1, in which for convenience, $(x_1, y_1, z_1) = \left(0, 0, \left[(r^*)^2 - (\frac{2}{3} r^* \sin 60°)^2\right]^{\frac{1}{2}}\right)$, $(x_2, y_2, z_2) = \left(0, \frac{2}{3} r^* \sin 60°, 0\right)$, $(x_3, y_3, z_3) = \left(\frac{1}{2} r^*, -\frac{1}{3} r^* \sin 60°, 0\right)$, $(x_4, y_4, z_4) = \left(-\frac{1}{2} r^*, -\frac{1}{3} r^* \sin 60°, 0\right)$.

For this arrangement, P_2, P_3, P_4 are in the XY plane and are equidistant from the origin, while P_1 lies on the Z-axis.

To derive a formula for the melting point of a solid, we begin by studying, in particular, copper. For this purpose, note first that a Lennard-Jones 6-12 potential

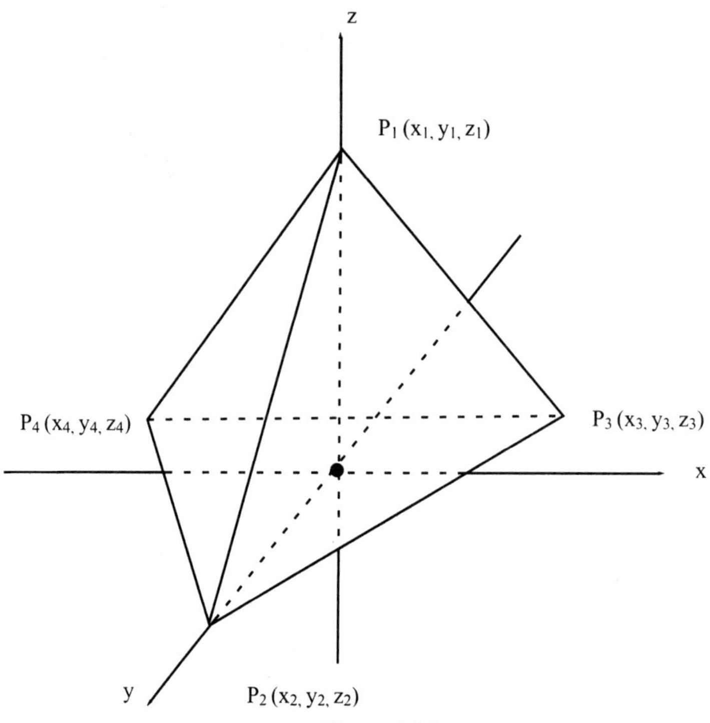

Figure 16.1.

for copper is (Greenspan (1989b)):

$$\phi(r) = -\left(\frac{1.398068}{r^6}\right) 10^{-10} + \left(\frac{1.55104}{r^{12}}\right) 10^{-8} \ (\text{erg}) \tag{16.1}$$

In dynes, the force $\vec{F}$ then has magnitude F given by

$$F = -\frac{(8.388408)}{r^7} 10^{-2} + \left(\frac{18.61248}{r^{13}}\right), \tag{16.2}$$

from which it follows readily that $r^* = 2.460486$Å. The Newtonian equations of
motion for P_i are then

$$m_i \vec{a}_i = \sum_{\substack{j=1 \\ j \neq i}}^{4} \left[-\frac{8.388408}{r_{ij}^7} (10^{-2}) + \left(\frac{18.61248}{r_{ij}^{13}}\right)\right] \frac{\vec{r}_{ji}}{r_{ij}} \left(\frac{\text{gr cm}}{\text{sec}^2}\right) \tag{16.3}$$

in which $\vec{r}_{ji}$ is the vector from P_j to P_i and $r_{ij} = \|\vec{r}_{ij}\|$. Since the mass of copper atom is $(1.0542)10^{-22}$ gr, equation (16.3) is equivalent to

$$(1.0542)10^{-22}\vec{a}_i = \sum_{\substack{j=1 \\ j \neq i}}^{4} \left[-\frac{(8.388408)}{r_{ij}^7}(10^{-2}) + \left(\frac{18.61248}{r_{ij}^{13}}\right) \right] \frac{\vec{r}_{ji}}{r_{ij}} \left(\frac{10^8 \text{\AA}}{\text{sec}^2}\right),$$

or

$$\frac{d^2\vec{r}_i}{dt^2} = \sum_{\substack{j=1 \\ j \neq i}}^{4} \left[-\frac{7.957132}{r_{ij}^7}10^{28} + \frac{17.65555}{r_{ij}^{13}}10^{30} \right] \frac{\vec{r}_{ji}}{r_{ij}} \qquad i = 1, 2, 3, 4 \quad (16.4)$$

To solve the system (16.4) efficiently, we make the time transformation

$$T = 10^{15}t \text{ sec},$$

from which it follows that

$$v_i = \frac{dr_i}{dt} = \frac{dr_i}{dT} \cdot \frac{dT}{dt} = \frac{dr_i}{dT} \cdot 10^{15} = V_i 10^{15} \qquad (16.5)$$

$$\frac{d^2r_i}{dt^2} = \frac{dv_i}{dt} = \frac{d^2r_i}{dT^2} \cdot 10^{30} \qquad (16.6)$$

and (16.4) reduces to

$$\frac{d^2\vec{r}_i}{dT^2} = \sum_{\substack{j=1 \\ j \neq i}}^{4} \left[-\frac{.07957132}{r_{ij}^7} + \frac{17.65555}{r_{ij}^{13}} \right] \frac{\vec{r}_{ji}}{r_{ij}}. \qquad (16.7)$$

For initial data we have

$$(x_1, y_1, z_1) = (0, 0, 2.008978)$$

$$(x_2, y_2, z_2) = (0, 1.420562, 0)$$

$$(x_3, y_3, z_3) = (1.230243, -0.7102811, 0)$$

$$(x_4, y_4, z_4) = (-1.230243, -0.7102811, 0)$$

and choose

$$\vec{V}_1 = (0, 0, -V_z), \qquad \vec{V}_2 = \vec{V}_3 = \vec{V}_4 = 0.$$

We will determine the minimum value V_z for which P_1 passes through the plane of P_2, P_3, P_4. Intuitively, such behavior is fluid like and should enable one to characterize the melting transition.

Beginning with $V_z = 0.1, 0.2, 0.3, 0.4, \ldots, 1.0$ and running each such choice for 750,000 time steps using the leap frog formulas with $\Delta T = 0.0001$ it is found that $0.2 < V_z < 0.3$. Refining V_z then to $V_z = 0.20, 0.21, 0.22, \ldots, 0.30$ and running each case again, it is found that $0.20 < V_z < 0.21$. Continuing in the indicated fashion, it is found that, to five decimal places, $V_z = 0.20380$. Thus, from (16.5),

$$v_z = (0.20380)10^{15}\text{Å}/s = (0.20380)10^7 \frac{\text{cm}}{\text{sec}} \, .$$

If $\bar{v}$ is the initial speed of P_1 *relative* to the mass center of the system, then $\bar{v} = \frac{3}{4} v_z = (0.15285)10^7$ cm/sec. We now define the melting point T_0 of copper in °K by

$$T_0 = C \left(\frac{1}{2} m\bar{v}^2 \right) , \tag{16.8}$$

in which C is a constant which is determined as follows. Since the mass of a copper atom is $(1.0542)10^{-22}$gr and its melting point is 1357 °K, equation (16.8) implies

$$1357 = C \left(\frac{1}{2} \right) (1.0542)10^{-22}(0.15285)^2 10^{14} \, ,$$

so that

$$C = (1.101935)10^{13}$$

but, to 0.2% one finds

$$C = \left(\frac{h}{6} \right) 10^{40} , \tag{16.9}$$

in which h is Planck's constant $(6.6251)10^{-27}$.

From the results above, we propose that the general formula for the melting point T_0 of *any* system is

$$T_0 = \left(\frac{h}{12} m\bar{v}^2 \right) 10^{40}\text{K} \tag{16.10}$$

and proceed to examine the applicability of this formula to other atomic species.

16.3 Noble Gas Calculations

For atomic interaction, potential formulas are available primarily for helium and the noble gases (Hirschfelder, Curtiss and Bird (1965)). We will direct attention in this section to the noble gases. It may be noted immediately that the formula (16.1) for copper was derived by a least square fit of an available dataset and only a limited number of such datasets seem to be available. As we turn attention to the noble gases, however, difficult problems of choice result immediately. Not only are there structurally different potential formulas available, like those of Buckingham, Corner, and Lennard-Jones, but each particular form may have a variety of parameter values available. For example, for argon, there are at least five different parameter values available for the general Lennard-Jones potential

$$\phi(r) = 4\epsilon \left[\left(\frac{\sigma}{r}\right)^{12} - \left(\frac{\sigma}{r}\right)^{6} \right]. \tag{16.11}$$

In the present section we will limit attention by considering only Lennard-Jones potentials and will consider exactly three different parameter sets for each noble gas. Using the Boltzmann constant $k = (1.38055)10^{-16}$, we have recorded in Table 16.1 the resulting parameters for (16.11) for each noble gas. Also listed are the respective masses, equilibrium distances r^* and minimum calculated values $V_{\tilde{z}}$. The experimental melting point (Dean (1985)) and theoretical melting point T_0, from (16.10), are recorded in the final columns of Table 16.1. These are given in °C because the experimental results have been determined in °C.

The calculated values T_0 in Table 16.1 indicate quite clearly that they are sensitive to the choices of ϵ and σ. Thus, the method requires highly accurate values of ϵ and σ for applicability. Nevertheless, for each noble gas, at least one value of T_0 is very accurate. In this connection, however, there is one minor deception in the table. The third case for krypton is not one available in the literature. Indeed, for the first two cases, the values of σ are essentially the same. It is interesting that different researchers have found the same σ, but have deduced such different values of ϵ. Our third case is then an artificial case in which we chose the same σ as in the first two cases but merely averaged the ϵ values, to yield the correct T_0.

Table 16.1   Noble Gas Calculations

Noble Gas	ϵ ($\times 10^{-16}$ erg)	σ (Å)	r^* (Å)	Mass ($\times 10^{-24}$ gr)	Computed $V_{\tilde{z}}$	Exp Melt. Pt. (°C)	T_0 (°C)
Ne	+49.14758	2.749	3.085648	33.47954	0.04804	-249	-249
	+48.18120	2.78	3.120444		0.04764		-249
	×49.2856	2.789	3.130547		0.04815		-249
Ar	+165.3899	3.405	3.821983	66.27881	0.06240	-189	-193
	+168.4271	3.40	3.816371		0.06292		-191
	×171.188	3.418	3.836575		0.06342		-190
Kr	+236.07405	3.60	4.040863	139.0348	0.05251	-157	-154
	+218.1269	3.597	4.037496		0.05066		-162
	•227.1005	3.60	4.040863		0.05150		-158
Xe	+305.1016	4.100	4.602094	217.8434	0.04881	-112	-112
	+299.5794	3.963	4.448317		0.04822		-116
	×316.1460	4.055	4.551584		0.04952		-107

[+] Parameters determined from second viral coefficients.
[×] Parameters determined from viscosity.
[•] See remark about this calculation.

Table 16.2   Helium Calculations

Potential	$r^*(\text{Å})$	Computed V_z	Exp. Melt. Pt. (°C)	T_0 (°C)
Lennard-Jones	2.869013	0.05683	-272.2	-266.34
Rosen-	0.9795108	0.002379	-272.2	-272.99
Marginau-Page	3.1835732	0.008665		-272.84
Slater-Kirkwood	2.4257225	0.006700	-272.2	-272.91
	2.94305228	0.008010		-272.87

16.4 Helium (26 atm)

Calculation of the melting point of helium requires special considerations. There are two reasons for this. First, though potential functions are available, the differentiation of these functions for dynamical calculations *magnifies* errors. Differentiation simply does not have the stable character of integration. Second, in degrees Kelvin, the experimental melting point is 0.8°K, which is so close to zero that numerical accuracy, after differentiation is introduced, is not easily achievable. In this section, then we will consider three structurally different potentials (Hirschfelder, Curtiss and Bird (1965)) and compare the results for each. These potentials are

$$\phi(r) = 4(14.10922)10^{-16}\left[\left(\frac{2.556}{4}\right)^2 - \left(\frac{2.556}{r}\right)^6\right]\text{erg}. \qquad (16.12)$$

(Lennard-Jones)

$$\phi(r) = \left[925e^{-4.40r} - 560e^{-5.33r} - \frac{1.39}{r^6} - \frac{3.0}{r^8}\right]10^{-12}\text{erg}. \qquad (16.13)$$

(Rosen-Marginau-Page)

$$\phi(r) = \begin{cases} \left[770e^{-4.60r} - \frac{1.49}{r^6}\right]10^{-12}\text{ erg}, & \text{if } r \leq 2.61 \\ \left[977e^{-4.60r} - \frac{1.5}{r^6} - \frac{2.51}{r^8}\right]10^{-12}\text{ erg}, & \text{if } r > 2.61 \end{cases} \qquad (16.14)$$

(Slater-Kirkwood)

Recall also that the mass of helium is $(6.64082)10^{-24}$ gr.

The computation using the Lennard-Jones potential follows in the fashion described in Sect. 16.3 and, as recorded in Table 16.2, yields $r^* = 2.869013$, $V_z = 0.05683$, $T_0 = -266.34$°C. The other two potentials require more extensive considerations, so let us consider next the Rosen-Marginau-Page potential in detail.

From the Rosen-Marginau-Page potential (16.13), it follows that

$$F = \left[4070e^{-4.40r} - 2984e^{-5.33r} - \frac{8.34}{r^7} - \frac{24.0}{r^9} \right] 10^{-4} \left(\frac{\text{gr } A}{\text{sec}^2} \right).$$

From the dynamical equation

$$(6.64082)10^{-24}a = F,$$

one finds

$$\frac{d^2r}{dt^2} = \left[(612.87612)e^{-4.40r} - 449.46257e^{-5.33r} \right.$$
$$\left. - \frac{1.255869}{r^7} - \frac{3.6140115}{r^9} \right] 10^{20} \frac{A}{\text{sec}^2}. \tag{16.15}$$

To determine r^* from (16.15), one must solve the transcendental equation

$$(612.87612)e^{-4.40r} - 449.46257e^{-5.33r} - \frac{1.255869}{r^7} - \frac{3.6140115}{r^9} = 0. \tag{16.16}$$

Interestingly enough, equation (16.16) has *two* solutions, namely $r^* = 0.97951080$ and $r^* = 3.1835732$.

Transforming (16.16) by $T = 10^{15}t$ reduces (16.15) to

$$\frac{d^2r}{dT^2} = \left[(612.87612)e^{-4.40r} - 449.46257e^{-5.33r} \right.$$
$$\left. - \frac{1.255869}{r^7} - \frac{3.6140115}{r^9} \right] 10^{-10}. \tag{16.17}$$

Applying the leap frog formulas to (16.17) with $\Delta T = 0.0004$ using $r^* = 0.97951080$ yields a minimum $V_z = 0.002379$ and, from (16.10), $T_0 = -272.99°C$, as recorded in Table 16.2. For $r^* = 3.1835732$, one finds a minimum $V_z = 0.008665$ and $T_0 = -272.84°C$.

Consider now the Slater-Kirkwood potential. It can be handled in a fashion entirely analogous to the Rosen-Marginau-Page potential as follows. From (16.14), the dynamical equations are

$$\frac{d^2r}{dT^2} = \left[533.36787e^{-4.60r} - \frac{1.3462193}{r^7} \right] 10^{-10}, \qquad r \le 2.61 \tag{16.18}$$

$$\frac{d^2r}{dT^2} = \left[676.75377e^{-4.60r} - \frac{1.3552543}{r^7} - \frac{3.023723}{r^9} \right] 10^{-11},$$
$$r > 2.61. \tag{16.19}$$

Now, the right-hand side of (16.18) has two values r^* which reduce it to zero. However, only one of the two is in the range $r \le 2.61$, and it is $r^* = 2.4257225$.

Similarly, for (16.19), there are two values of r^*, but only one is in the range $r \geq 2.61$, namely, $r^* = 2.9430528$. For $r^* = 2.4257225$, the minimum $V_{\c}$ is 0.006700 and $T_0 = -272.91°C$, while for $r^* = 2.94305228$, the minimum value $V_{\c}$ is 0.008010 and $T_0 = -272.87°C$, as recorded in Table 16.2.

16.5 Homogeneous, Diatomic Molecular Solids

We turn next to solids which are composed of homogeneous, diatomic molecules. For these, we will consider all available cases for which there exist at least two Lennard-Jones formulas of type (16.1) and will apply the methodology of Sect. 16.3. The molecules considered are H_2, D_2, N_2, O_2, and Cl_2 (Hirschfelder, Curtiss and Bird (1965)).

Using the same tabular notations as in Table 16.1, we have listed the parameter values and computational results in Table 16.3.

Again, except for the Cl_2 case, each molecule has at least one related parameter set for which the calculation yields excellent results. For Cl_2 note that only two parameter sets were available, neither of which was derived from quantum mechanical considerations. Moreover, even in the Cl_2 case, one of the results is in error by only 4%.

Table 16.3   Diatomic Molecule Calculations

Molecule	ϵ $(\times 10^{-16}$ erg)	σ (Å)	r^*	Mass $(10^{-24}$ gr)	Computed V_z	Exp Melt. Pt. (°C)	T_0 (°C)
H_2	$^+$51.08035	2.928	3.28656888	3.3448	0.14737	-259.19	-250.4
	$^+$45.97232	2.968	3.33146736		0.14004		-252.6
	$^\times$52.46090	2.915	3.27197687		0.14928		-249.9
	$^+$40.31206	2.87	3.22146608		0.13125		-255.1
D_2	$^+$51.08035	2.928	3.28656888	6.6896	0.10529	-252.89	-249.97
	$^+$42.935105	2.87	3.22146608		0.9676		-253.55
	$^\times$54.255615	2.948	3.30901812		0.10842		-248.58
N_2	$^+$132.3948	3.71	4.1643320	46.5028	0.06676	-209.86	-208.6
	$^+$131.2213	3.698	4.15086465		0.06650		-209.1
	$^\times$126.3203	3.681	4.13178280		0.06645		-209.2
	$^\times$110.1679	3.749	4.20811021		0.06681		-208.5
O_2	$^+$162.9049	3.46	3.88371869	53.1186	0.06855	-218.4	-200.2
	$^+$162.2146	3.58	4.01841413		0.06857		-205.6
	$^\times$156.00215	3.433	3.85341221		0.06715		-213.6
	$^\times$121.4884	3.541	3.97463811		0.05992		-188.1
Cl_2	$^\times$492.8564	4.115	4.61893133	117.7037	0.08047	-100.98	-36.3
	$^\times$354.80135	4.4	4.93883301		0.06946		-96.64

$^+$Parameters determined from second viral coefficients.
$^\times$Parameters determined from viscosity.

17

Special Relativistic Motion

17.1 Introduction

Deterministic particle motion can also be studied in a relativistic context. We will do this in the present chapter. However, only pertinent concepts from special relativity will be defined and applied. In addition, we will concentrate on the fundamental problem of motion in one space dimension.

In Newtonian mechanics one assumes that in making observations the speed of light is infinite. Thus, in Newtonian mechanics one assumes that any observed event is observed at the exact time it occurs. In fact, the speed of light is not infinite but is, approximately, 186,000 mile/sec. This constant will be denoted by c. That the time involved in making an observation can be significant is apparent in the following two cases:

a. when observing very distant objects, like distant galaxies, and

b. when observing objects moving close to the speed of light, like an electron which has been accelerated to $0.95c$.

The subject in which the finite speed of light is taken into account is called Relativity. Special Relativity uses the Euclidean formula to measure the distance between two points, while General Relativity uses a more general metric formula.

17.2 Inertial Frames

Consider two x axes, which at some time $t = 0$ are coincident and identical. Denote one by X and the second by X'. At the origin of each assume there are identical observers, each with identical and synchronized clocks. Assume finally that the X' axis is in motion relative to the X axis with constant speed u. The X

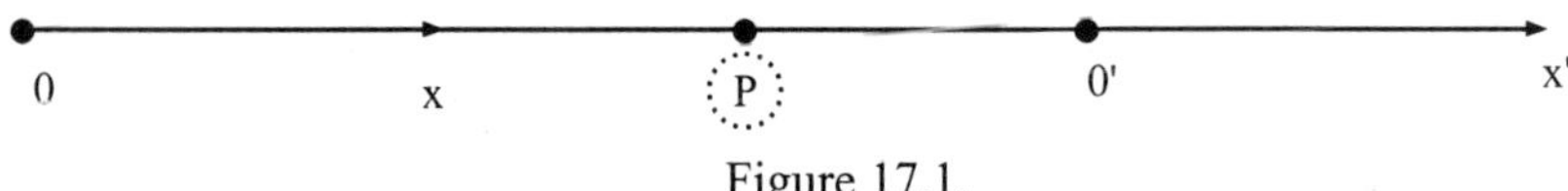

Figure 17.1.

and X' are called inertial systems, the X axis being called the lab frame and the X' axis being called the rocket frame. The term *rocket frame* is used to denote the importance of u being a large constant.

The Axiom of Continuity for inertial frames is as follows: All the laws of physics are the same in every inertial reference frame. This includes the invariance of all physical constants, including the speed of light.

17.3 The Lorentz Transformation

Consider now two inertial frames X and X' at some time $t > 0$, as shown in Fig. 17.1. Suppose the observers at O and O' both observe an event, like an exploding star, at P. Let P occur in the lab frame at position x and time t while in the rocket frame it occurs at position x' and time t'. Then (x, t) and (x', t') are called events because they incorporate knowledge of both position and time of occurrence. Taking into account the speed of light in making observations, H. A. Lorentz proved that the relationships between (x, t) and (x', t') are

$$x' = \frac{c(x - ut)}{(c^2 - u^2)^{\frac{1}{2}}}, \qquad t' = \frac{c^2 t - ux}{c(c^2 - u^2)^{\frac{1}{2}}}, \tag{17.1a}$$

or, equivalently,

$$x = \frac{c(x' + ut')}{(c^2 - u^2)^{\frac{1}{2}}}, \qquad t = \frac{c^2 t' + ux'}{c(c^2 - u^2)^{\frac{1}{2}}}. \tag{17.1b}$$

Equations (17.1) are called the Lorentz transformation.

To avoid singularities, we assume that $|u| < c$.

17.4 Rod Contraction and Time Dilation

The Lorentz formulas have interesting physical implications, two of which will be discussed now. The first is called the contraction of a moving rod and is described as follows.

Consider a thin rod which lies in the lab frame. Then, the rod's length is measured to be greatest when the rod is at rest relative to the lab frame observer.

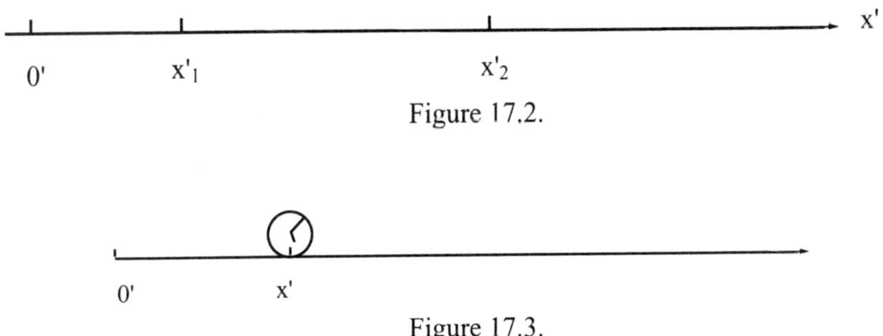

Figure 17.2.

Figure 17.3.

When it moves with speed u relative to the observer, its measured length is contracted by the factor $\sqrt{1 - \frac{u^2}{c^2}}$. To see this, let the rod lie at rest along the X' axes the rocket frame, as shown in Fig. 17.2.

Let its end points be x_2' and x_1', so that its length is $x_2' - x_1'$. Now, at time t, let the rod's length be measured by the lab frame observer. Then the end point coordinates x_2 and x_1 as observed in the lab frame are related to x_2', x_1' by

$$x_2' = \frac{c(x_2 - ut)}{(c^2 - u^2)^{\frac{1}{2}}}, \qquad x_1' = \frac{c(x_1 - ut)}{(c^2 - u^2)^{\frac{1}{2}}} \qquad (17.2)$$

Thus,

$$(x_2' - x_1') = \frac{c(x_2 - x_1)}{(c^2 - u^2)^{\frac{1}{2}}},$$

or

$$x_2 - x_1 = (x_2' - x_1')\left(\frac{c^2 - u^2}{c^2}\right)^{\frac{1}{2}} = (x_2' - x_1')\sqrt{1 - \frac{u^2}{c^2}}. \qquad (17.3)$$

However,

$$\sqrt{1 - \frac{u^2}{c^2}} < 1, \qquad (17.4)$$

which proves the contention, if one also observes that $\sqrt{1 - \frac{u^2}{c^2}} = 1$ if and only if $u = 0$.

Next observe that a clock is measured in the lab frame to go at its fastest rate when it is at rest relative to the lab frame observer. When the clock moves with speed u relative to the lab frame observer, its rate is measured to have slowed down by the factor $\sqrt{1 - \frac{u^2}{c^2}}$. If its rate of ticking slows down, time is dilated, that is, for example, the clock takes longer to measure an hour. To prove this, consider a clock which is at rest at a position x' the rocket frame, as shown in Fig. 17.3.

Consider two successive readings in the rocket frame at, say, t_1' and t_2'. The time interval in the rocket is $t_2' - t_1'$. To the lab observer, these times are recorded as t_1 and t_2, where

$$t_1 = \frac{c^2 t_1' + ux'}{c(c^2 - u^2)^{\frac{1}{2}}}, \qquad t_2 = \frac{c^2 t_2' + ux'}{c(c^2 - u^2)^{\frac{1}{2}}}. \tag{17.5}$$

However,

$$t_2 - t_1 = \frac{c^2(t_2' - t_1')}{c(c^2 - u^2)^{\frac{1}{2}}} = \frac{(t_2' - t_1')}{\sqrt{1 - \frac{u^2}{c^2}}}. \tag{17.6}$$

But

$$\frac{1}{\sqrt{1 - \frac{u^2}{c^2}}} > 1, \tag{17.7}$$

from which the assertion follows.

17.5 Relativistic Particle Motion

Consider now a particle P in motion in the lab frame. Then its velocity v and acceleration a are defined in the usual way

$$v = \frac{dx}{dt}, \qquad a = \frac{dv}{dt}. \tag{17.8}$$

By the Axiom of Continuity, one must have in the rocket

$$v' = \frac{dx'}{dt'}, \qquad a' = \frac{dv'}{dt'}. \tag{17.9}$$

With regard to (17.8) and (17.9), we assume $|v| < c$ and $|v'| < c$.
To relate v and v', we have from (17.1a)

$$v' = \frac{dx'}{dt'} = \frac{c(dx - u\,dt)}{(c^2 - u^2)^{\frac{1}{2}}} \div \left(\frac{c^2 dt - u\,dx}{c(c^2 - u^2)^{\frac{1}{2}}} \right) = \frac{c^2(dx - u\,dt)}{(c^2 dt - u\,dx)},$$

so that

$$v' = \frac{c^2(v - u)}{c^2 - uv}. \tag{17.10}$$

Equivalently,

$$v = \frac{c^2(v' + u)}{c^2 + uv'}. \tag{17.11}$$

Similarly, the relationship between a and a' is found to be

$$a' = \frac{c^3(c^2 - u^2)^{\frac{3}{2}}}{(c^2 - uv)^3} a, \tag{17.12}$$

or equivalently, by

$$a = \frac{c^3(c^2 - u^2)^{\frac{3}{2}}}{(c^2 + uv')^3}\, a' \qquad .(17.13)$$

17.6 Covariance

By covariance, one means that the structure of the dynamical equations associated with a physical formulation is invariant under fundamental coordinate transformations. In special relativity, this means under the Lorentz transformation.

Around 1900 it was shown that Newton's dynamical equation was not invariant under the Lorentz transformation, whereas Maxwell's equations, that is, the equations of electromagnetics, were. The question arose, then, as to what dynamical equation for particle motion was covariant under the Lorentz transformation, and Einstein showed that if we assume that mass varies with speed, then with only a slight modification of Newton's equation the result was a covariant dynamical equation. Let us prove this result first, since it is essential for an understanding of later discussions.

THEOREM 17.7
Let a particle P be in motion along the X axis in the lab frame and along the X′ axis in the rocket frame. In the lab frame, let the mass m of P be given by

$$m = \frac{cm_0}{(c^2 - v^2)^{\frac{1}{2}}} \qquad (17.14)$$

where m_0 is a positive constant called the rest mass of P and v is the speed of P in the lab. In the rocket frame, let the mass m′ of P be given by

$$m' = \frac{cm_0}{(c^2 - (v')^2)^{\frac{1}{2}}}, \qquad (17.15)$$

where m_0 is the same constant as in (17.14) and v′ is the speed of P in the rocket. Let a force F be applied to P in the lab. In rocket coordinates, denote the force by F′, so that

$$F = F'.$$

Then, if in the lab, the equation of motion of P is given by

$$F = \frac{d}{dt}(mv), \qquad (17.16)$$

it follows that the equation of motion in the rocket is

$$F' = \frac{d}{dt'}(m'v'). \qquad .(17.17)$$

PROOF From (17.14) and (17.16),

$$F = v\,\frac{dm}{dt} + m\,\frac{dv}{dt} = v\left[\frac{\left(-\frac{1}{2}(cm_0)\right)}{(c^2 - v^2)^{\frac{3}{2}}}(-2va)\right] + ma = \frac{v^2 ma}{c^2 - v^2} + ma\,,$$

so that

$$F = \left(\frac{c^2}{c^2 - v^2}\right)ma \tag{17.18}$$

From (17.15) and (17.17), then, we must have

$$F' = \left(\frac{c^2}{c^2 - (v')^2}\right)m'a' \tag{17.19}$$

Since $F = F'$, the proof will follow if we can establish the identity

$$ma\left(\frac{c^2}{c^2 - v^2}\right) \equiv m'a'\left(\frac{c^2}{c^2 - (v')^2}\right)\,,$$

or, equivalently,

$$\frac{ma}{c^2 - v^2} \equiv \frac{m'a'}{c^2 - (v')^2} \tag{17.20}$$

However, substitution of (17.10), (17.12) and (17.15) into the right side of (17.20) yields, remarkably, that the identity is valid and the theorem is proved. $\square$

17.7 Relativistic Motion

Let us consider now a particle P which moves on the X axis in the lab frame. Assume also that the force F on P is one whose magnitude depends only on the x coordinate of P. Then, let

$$F = f(x)\,. \tag{17.21}$$

Assume that initially, i.e., at time $t = 0$, P is at x_0 and has speed v_0. Then the equation of motion of P in the lab frame is

$$\frac{d}{dt}(mv) = f(x)\,. \tag{17.22}$$

From (17.18)

$$\frac{c^2}{(c^2 - v^2)}\,ma = f(x) \tag{17.23}$$

or

$$c^2 m\ddot{x} = f(x)(c^2 - \dot{x}^2)$$

From (17.14), this can be reduced to

$$c^3 \ddot{x} m_0 = f(x)(c^2 - \dot{x}^2)^{\frac{3}{2}},$$

so that, finally,

$$\ddot{x} - \frac{f(x)}{c^3 m_0}(c^2 - \dot{x}^2)^{\frac{3}{2}} = 0 \tag{17.24}$$

is the differential equation one has to solve in the lab frame, given the initial data

$$x(0) = x_0, \qquad \dot{x}(0) = v_0. \tag{17.25}$$

In general, (17.23) cannot be solved in closed form, so that the observer in the lab frame must now introduce a computer to approximate the solution. However, the observer in rocket frame also observes the motion of P, but in his coordinate system. His equation and initial conditions are found by applying (17.7), (17.11) and (17.13) to (17.24) and (17.25). Thus, he too will be confronted with a differential equations problem which requires computational methodology, so a computer identical to that in the lab is introduced also into the rocket.

A fundamental problem in preserving the physics of special relativity then arises in how the two observers should approximate the solutions of their initial value problems. Their differential equations are covariant under the Lorentz transformation. To preserve the physics, if they use difference approximations, then the difference approximations of these differential equations should also be covariant under the Lorentz transformation. We will show next how this can be accomplished.

17.8 Numerical Methodology

In the lab, let $\Delta t > 0$ and $t_k = k\Delta t$. At time t_k, let P be at x_k in the lab. Then, in the rocket P will be at x'_k at time t'_k, where

$$x'_k = \frac{c(x_k - ut_k)}{(c^2 - u^2)^{\frac{1}{2}}}, \qquad t'_k = \frac{c^2 t_k - ux_k}{c(c^2 - u^2)^{\frac{1}{2}}} \tag{17.26}$$

or equivalently,

$$x_k = \frac{c(x'_k + ut'_k)}{(c^2 - u^2)^{\frac{1}{2}}}, \qquad t_k = \frac{c^2 t'_k + ux'_k}{c(c^2 - u^2)^{\frac{1}{2}}}. \tag{17.27}$$

The formulas (17.26) and (17.27) are valid because they are merely special cases of (17.1), that is, they result from the particular choices $x = x_k$ and $t = t_k$.

The concepts of velocity and acceleration are now approximated by the following formulas. At t_k in the lab, let

$$v_k = \frac{\Delta x_k}{\Delta t_k} = \frac{x_{k+1} - x_k}{t_{k+1} - t_k}, \qquad a_k = \frac{\Delta v_k}{\Delta t_k} = \frac{v_{k+1} - v_k}{t_{k+1} - t_k}. \tag{17.28}$$

At t'_k in the rocket, let

$$v'_k = \frac{\Delta x'_k}{\Delta t'_k} = \frac{x'_{k+1} - x'_k}{t'_{k+1} - t'_k}, \qquad a'_k = \frac{\Delta v'_k}{\Delta t'_k} = \frac{v'_{k+1} - v'_k}{t'_{k+1} - t'_k}. \qquad (17.29)$$

Then, corresponding to (17.10)–(17.13), one has by direct substitution that

$$v'_k = \frac{x^2(v_k - u)}{c^2 - uv_k}, \qquad v_k = \frac{c^2(v'_k + u)}{c^2 + uv'_k} \qquad (17.30)$$

$$a'_k = \frac{c^3(c^2 - u^2)^{\frac{3}{2}}}{(c^2 - uv_k)^2(c^2 - uv_{k+1})} a_k, \qquad a_k = \frac{c^3(c^2 - u^2)^{\frac{3}{2}}}{(c^2 + uv'_k)^2(c^2 + uv'_{k+1})} a'_k. \qquad (17.31)$$

Of course, in the limit, (17.30) and (17.31) converge to (17.10)–(17.13). Our problem now is one of choosing an approximation to

$$F = \frac{d}{dt}(mv) \qquad (17.32)$$

in the lab which will transform covariantly into the rocket. The clue for this choice comes from (17.18), which is equivalent to (17.32). What we choose at t_k is the approximation

$$F_k = \frac{m(t_k)c^2}{\left[(c^2 - v_k^2)(c^2 - v_{k+1}^2)\right]^{\frac{1}{2}}} a_k, \qquad m(t_k) = \frac{cm_0}{\sqrt{c^2 - v_k^2}} \qquad (17.33)$$

Note first that, again, in the limit, (17.33) converges to (17.18). What we must prove is that if $F_k = F'_k$, and if in the rocket

$$m'(t'_k) = \frac{cm_0}{\sqrt{c^2 - (v'_k)^2}}, \qquad (17.34)$$

then in the rocket

$$F'_k = \frac{m'(t'_k)c^2}{\left[(c^2 - (v'_k)^2)(c^2 - (v'_{k+1})^2)\right]^{\frac{1}{2}}} a'_k. \qquad (17.35)$$

THEOREM 17.8
If $F_k = F'_k$, then (17.33), (17.34) imply (17.35).

PROOF The proof is entirely analogous to that of Theorem 17.1, but uses (17.26)–(17.31). $\square$

We will show now how, in the lab, (17.33) is applied to generate the numerical solution of an oscillator initial value problem.

For simplicity in further discussions we will use the normalization constants $m_0 = c = 1$. Then recursion formulas for the motion of an oscillator are, in the lab, from (17.28) and (17.33).

$$x_{k+1} = x_k + (t_{k+1} - t_k)v_k \tag{17.36}$$

$$v_{k+1} = v_k + (t_{k+1} - t_k)(1 - v_k^2)(1 - v_{k+1}^2)^{\frac{1}{2}} F_k. \tag{17.37}$$

Without loss of generality, we consider now only $k = 0$, since each recursive step is essentially the same except for subscript. Hence,

$$x_1 = x_0 + (t_1 - t_0)v_0 \tag{17.38}$$

$$v_1 = v_0 + (t_1 - t_0)(1 - v_0^2)(1 - v_1^2)^{\frac{1}{2}} F_0. \tag{17.39}$$

The problem in applying (17.38)–(17.39) is that (17.39) is implicit in v_1. Let us show, however, that (17.39) can be solved explicitly and uniquely for v_1. Set $A = (t_1 - t_0)(1 - v_0^2)F$. Then (17.39) reduces to

$$(v_1 - v_0) = A(1 - v_1^2)^{\frac{1}{2}}. \tag{17.40}$$

Now, $1 - v_0^2 > 0$, $1 - v_1^2 > 0$ and $1 - u^2 > 0$, by the assumption that no speed is greater than c, which has been normalized to unity. Thus, if $F = 0$, then $A = 0$. Moreover, if $F > 0$ then $A > 0$, while if $F < 0$ then $A < 0$. Thus, A has the same sign as F. Now, if $A = 0$, then $v_1 = v_0$ and so v_1 is unique. Assume then that $A > 0$. Then, squaring both sides of (17.40) yields

$$v_1^2 - 2v_1 v_0 + v_0^2 = A^2(1 - v_1^2), \tag{17.41}$$

so that

$$v_1^2(1 + A^2) - 2v_0 v_1 + v_0^2 - A^2 = 0. \tag{17.42}$$

The solutions to this quadratic are

$$v_1 = \frac{v_0 \pm A\sqrt{1 - v_0^2 + A^2}}{1 + A^2}. \tag{17.43}$$

However, substitution of (17.43) into (17.40) reveals that only

$$v_1 = \frac{v_0 + A\sqrt{1 - v_0^2 + A^2}}{1 + A^2} \tag{17.44}$$

is a root of (17.43). The same result follows in the case $A < 0$. Thus, (17.36) and (17.37) can be replaced by the explicit formulas

$$x_{k+1} = x_k + (t_{k+1} - t_k)v_k \tag{17.45}$$

$$v_{k+1} = \frac{v_k + (t_{k+1} - t_k)(1 - v_k^2)F_k\sqrt{1 - v_k^2 + (t_{k+1} - t_k)^2(1 - v_k^2)^2 F_k^2}}{1 + (t_{k+1} - t_k)^2(1 - v_k^2)^2 F_k^2}$$

$$\tag{17.46}$$

17.9 Relativistic Harmonic Oscillation

For the usual definition of a harmonic oscillator in Newtonian mechanics, one assumes that $f(x)$ in (17.21) is given by $f(x) = -K^2 x$, where K is a nonzero constant. We will assume that this same choice of $f(x)$ in (17.22) defines a relativistic harmonic oscillator and examine its motion in the lab frame for given initial data.

We now set $K = 1$, in addition to $m_0 = c = 1$. Then (17.24) reduces to

$$\ddot{x} - x(1 - \dot{x}^2)^{\frac{3}{2}} = 0 \tag{17.47}$$

To generate a numerical solution, observe now that

$$x_{k+1} = x_k + (\Delta t_k)v_k , \qquad k = 0, 1, \ldots . \tag{17.48}$$

$$v_{k+1} = \frac{v_k + (\Delta t_k)x_k(1 - v_k^2)^{\frac{3}{2}} \left[1 + x_k^2(\Delta t_k)^2(1 - v_k^2)\right]^{\frac{1}{2}}}{1 + x_k^2(\Delta t_k)^2(1 - v_k^2)^2} , \tag{17.49}$$

$$k = 0, 1, 2, \ldots .$$

Let us assume now that $x(0) = x_0 = 0$ and examine the results for various values of v_0 in the range $0 < v_0 < 1$. In particular, this is done for 30,000 time steps with $\Delta t = 0.0001$ for each of the cases $v_0 = 0.001, 0.01, 0.05, 0.1, 0.3, 0.5, 0.7, 0.9$. Then Fig. 17.4 shows the amplitude and period of the first complete oscillation for the case $v_0 = 0.001$. For such a relatively low velocity, the oscillator should behave like a Newtonian oscillator and, indeed, this is the case, with the amplitude being 0.001 and, to two decimal places, the period being 6.28 ($\sim 2\pi$). Subsequent motion of this oscillator continues to show almost no change in amplitude or period. At the other extreme, Fig. 17.5 shows the motion for $v_0 = 0.9$, which is relatively close to the speed of light. To two decimal places, the amplitude of the first oscillation is 1.61 while the period is 8.88. These results are distinctly non-Newtonian, and, to 30,000 time steps, these results remain constant to two decimal places but do show small increments in the third decimal place. Finally, in Fig. 17.6 is shown how the amplitude of the relativistic harmonic oscillator deviates from that of the Newtonian harmonic oscillator with increasing v_0.

17.10 Computational Covariance

The discussions in Sects. 17.8 and 17.9 are now shown to be consistent in that computations performed in the lab and corresponding computations done in the rocket are themselves related by the Lorentz transformation. To do this, we proceed as follows.

In the lab the computations are defined by

$$x_{k+1} = x_k + v_k(\Delta t_k) \tag{17.50}$$

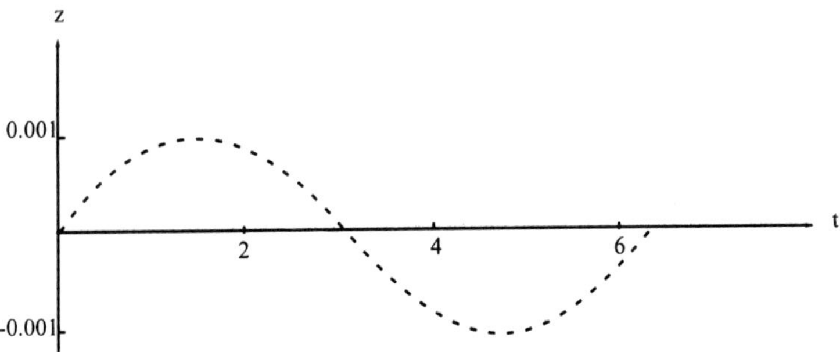

Figure 17.4.

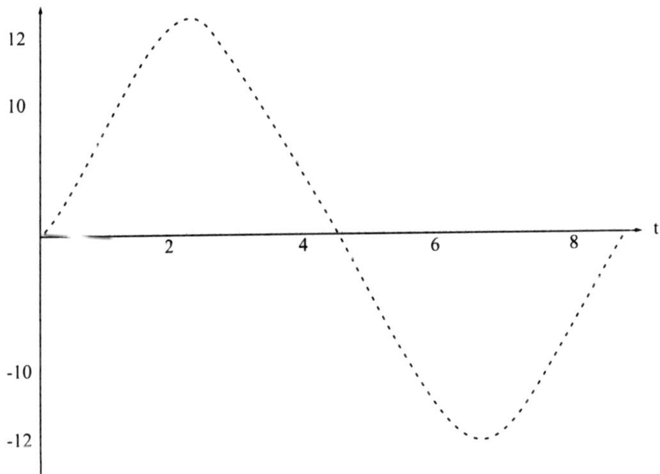

Figure 17.5.

$$v_{k+1} = \frac{v_k + (\Delta t_k)(1 - v_k^2)F_k\sqrt{1 - v_k^2 + (\Delta t_k)^2(1 - v_k^2)^2 F_k^2}}{1 + (\Delta t_k)^2(1 - v_k^2)^2 F_k^2}, \qquad (17.51)$$

in which

$$\Delta t_k = t_{k+1} - t_k$$

$$F_k = f(x_k).$$

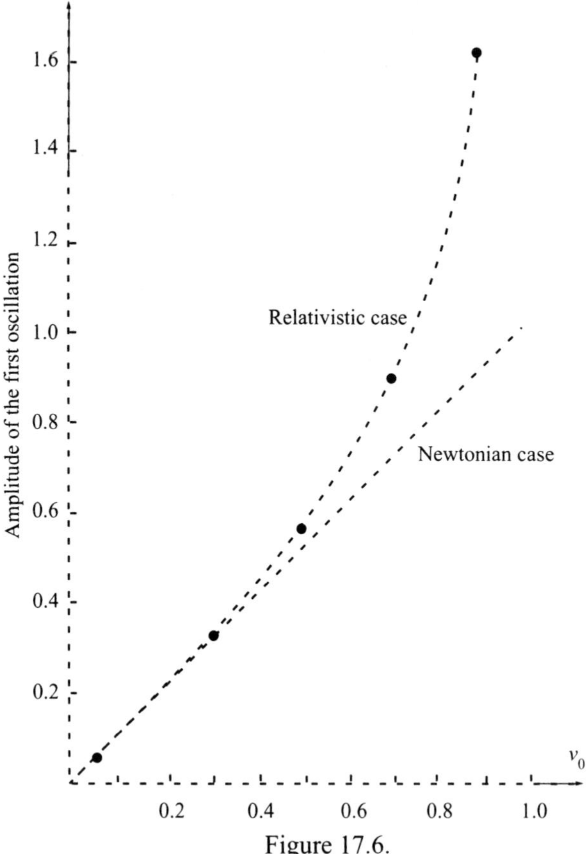

Figure 17.6.

By the Axiom of Continuity, calculations in the rocket would use

$$x'_{k+1} = x'_k + v'_k(\Delta t'_k) \tag{17.52}$$

$$v'_{k+1} = \frac{v'_k + (\Delta t'_k)(1 - v'^2_k)F'_k\sqrt{1 - (v'_k)^2 + (\Delta t'_k)^2(1 - v'^2_k)^2 F'^2_k}}{1 + (\Delta t'_k)^2(1 - v'^2_k)^2 F'^2_k}, \tag{17.53}$$

in which

$$\Delta t'_k = t'_{k+1} - t'_k$$

$$F'_k = f\left(\frac{x'_k + ut'_k}{(1 - u^2)^{\frac{1}{2}}}\right) = f(x_k) = F_k$$

We want to prove that

$$x'_{k+1} = \frac{x_{k+1} - ut_{k+1}}{(1 - u^2)^{\frac{1}{2}}} \tag{17.54}$$

$$v'_{k+1} = \frac{v_{k+1} - u}{1 - u v_{k+1}} \tag{17.55}$$

which will establish that computational results in the lab and the rocket are related by the Lorentz transformation.

We assume that x_0, x'_0, v_0, v'_0, the given initial data, are related by the Lorentz transformation. We will then prove (17.54), (17.55) for $k = 0$. This will be sufficient, by induction, to establish (17.54) and (17.55). Thus,

$$x'_0 = \frac{x_0 - u t_0}{(1 - u^2)^{\frac{1}{2}}}.$$

$$v'_0 = \frac{v_0 - u}{1 - u v_0}$$

Now,

$$x'_1 = x'_0 + v'_0(\Delta t'_0). \tag{17.56}$$

Let $\bar{x}_1$ correspond to x'_1 by the Lorentz transformation, so that

$$x'_1 = \frac{\bar{x}_1 - u t_1}{(1 - u^2)^{\frac{1}{2}}}.$$

We want to show now that $\bar{x}_1 = x_1$.

We know in addition that

$$t'_0 = \frac{t_0 - u x_0}{(1 - u^2)^{\frac{1}{2}}}, \qquad t'_1 = \frac{t_1 - u \bar{x}_1}{(1 - u^2)^{\frac{1}{2}}}.$$

Thus,

$$\frac{\bar{x}_1 - u t_1}{(1 - u^2)^{\frac{1}{2}}} = x'_0 + (v'_0)(\Delta t'_0)$$

$$= \frac{x_0 - u t_0}{(1 - u^2)^{\frac{1}{2}}} + \left(\frac{v_0 - u}{1 - u v_0} \right) \left(\frac{t_1 - u \bar{x}_1}{(1 - u^2)^{\frac{1}{2}}} - \frac{t_0 - u x_0}{(1 - u^2)^{\frac{1}{2}}} \right)$$

Since $(1 - u^2) > 0$, it then follows that

$$\bar{x}_1 - u t_1 = x_0 - u t_0 + \frac{v_0 - u}{1 - u v_0}(t_1 - t_0 + u x_0 - u \bar{x}_1)$$

or

$$(\bar{x}_1 - u t_1)(1 - u v_0) = (x_0 - u t_0)(1 - u v_0) + (v_0 - u)(t_1 - t_0) + (v_0 - u)(u x_0 - u \bar{x}_1)$$

which simplifies to

$$(\bar{x}_1 - x_0)(1 - u^2) = v_0(1 - u^2)(t_1 - t_0).$$

Then

$$\bar{x}_1 - x_0 = v_0(t_1 - t_0)$$

so that

$$\bar{x}_1 = x_0 + v_0(t_1 - t_0) \,.$$

Thus, $\bar{x}_1 = x_1$.

Now, for $k = 0$,

$$v_1' = \frac{v_0' + (t_1' - t_0')(1 - v_0'^2)F'\sqrt{1 - (v_0')^2 + (t_1' - t_0')(1 - v_0'^2)^2(F')^2}}{1 + (t_1' - t_0')^2(1 - v_0'^2)^2(F')^2} \,. \tag{17.57}$$

Substitution of

$$t_1' = \frac{t_1 - ux_1}{(1 - u^2)^{\frac{1}{2}}} \,, \qquad t_0' = \frac{t_0 - ux_0}{(1 - u^2)^{\frac{1}{2}}}$$

$$F' = F = \frac{v_1 - v_0}{(t_1 - t_0)(1 - v_0^2)(1 - v_1^2)^{\frac{1}{2}}}$$

$$x_0' = \frac{x_0 - ut_0}{(1 - u^2)^{\frac{1}{2}}} \,, \qquad v_0' = \frac{v_0 - u}{1 - uv_0}$$

into (17.57) yields

$$v_1' = \frac{(v_1 - u)[(1 - v_0v_1)(1 - uv_0) + (v_1 - v_0)(u - v_0)]}{(1 - uv_1)[(1 - v_0v_1)(1 - uv_0) + (v_1 - v_0)(u - v_0)]} \,.$$

Finally, we will have the desired result

$$v_1' = \frac{v_1 - u}{1 - uv_1} \,,$$

provided the terms in the brackets, which are identical, are not zero. And we will
show this next.

Note first that since $|u| < 1$ and $|v| < 1$, then

$$|u - v| < 1 - uv \,.$$

This follows since

$$(1 - u^2)(1 - v^2) > 0$$

$$(u^2 - 1)(1 - v^2) < 0$$

$$u^2 + v^2 - 1 - u^2v^2 < 0$$

$$u^2 + v^2 < 1 + u^2v^2$$

$$u^2 - 2uv + v^2 < 1 + u^2v^2 - 2uv$$

$$(u - v)^2 < (1 - uv)^2$$

and

$$|u - v| < |1 - uv| = 1 - uv \,.$$

However,

$$(1 - v_0 v_1)(1 - u v_0) + (v_1 - v_0)(u - v_0) > 0$$

since

$$|v_1 - v_0| < 1 - v_0 v_1$$

$$|u - v_0| < 1 - u v_0,$$

so that

$$-(v_1 - v_0)(u - u_0) \le |v_1 - v_0| \cdot |u - u_0| < (1 - v_0 v_1)(1 - u v_0).$$

Hence

$$(1 - v_0 v_1)(1 - u v_0) + (v_1 - v_0)(u - v_0) > 0.$$

18

A Speculative Model of the Diatomic Molecular Bond

18.1 Introduction

The intimate relationship between wave length and energy levels enables one to use the steady state Schrödinger wave equation to deduce many molecular vibrational constants without actually simulating the vibrational motions themselves. However, simulation of vibrational motions by means of the nonsteady Schrödinger equation presents difficulties at the present time (Borman (1990), Polanyi (1987)).

In this chapter, we will explore the possibility of simulating vibrations of diatomic molecules by a classical approach, whose reasonableness has been established recently by Gell-Mann and Hartle (1993).

18.2 Classical Simulation of the Hydrogen Molecule

The inadequacy of Newtonian mechanics on the atomic and molecular levels is readily apparent if one attempts to simulate a vibrating, ground state, hydrogen molecule using only Coulombic forces. It will be instructive, however, to show this in detail in the present section.

Recall first that the ground state energy of H_2 is $(-5.1104)10^{-11}$ erg, the vibrational frequency of the protons is $(1.3)10^{14} Hz$, and the bond length is 0.74 Å (Herzberg (1965)).

In a ground state H_2 molecule, denote the electrons by P_1, P_3 and the protons by P_2, P_4. Classically, assume P_1, P_2, P_3, P_4 are point sources and that the only forces of interaction are Coulombic.

In cgs units, for $i = 1, 2, 3, 4$, and at any time t, let P_i be located at $\vec{r}_i = (x_i, y_i, z_i)$, have velocity $\vec{v}_i = (\dot{x}_i, \dot{y}_i, \dot{z}_i)$, and have acceleration $\vec{a}_i = (\ddot{x}_i, \ddot{y}_i, \ddot{z}_i)$.

Then the classical equations of motion for the P_i are

$$m_i \vec{a}_i = \sum_{\substack{j=1 \\ j \neq i}}^{4} \frac{e_i e_j}{r_{ij}^2} \frac{\vec{r}_{ji}}{r_{ij}}, \qquad i = 1, 2, 3, 4, \tag{18.1}$$

in which $\vec{r}_{ji}$ is the vector from P_j to P_i, $r_{ij} = \|\vec{r}_{ij}\|$, and

$$e_1 = e_3 = -e_2 = -e_4 = -(4.8028)10^{-10} \text{esu} \tag{18.2}$$

$$m_1 = m_3 = (9.1085)10^{-28} \text{gr} \tag{18.3}$$

$$m_2 = m_4 = (16724)10^{-28} \text{gr}. \tag{18.4}$$

For computational convenience, we now set $\vec{R}_i = (X_i, Y_i, Z_i)$ and make the transformations

$$\vec{R}_i = 10^{12} \vec{r}_i \tag{18.5}$$

$$T = 10^{22} t. \tag{18.6}$$

Then the system (18.1) of 12 equations in the 12 unknowns $x_i, y_i, z_i, i = 1, 2, 3, 4$ transforms readily into the following equivalent system:

$$\frac{d^2 X_1}{dT^2} = (2.5324576)\left(-\frac{X_1 - X_2}{R_{12}^3} + \frac{X_1 - X_3}{R_{13}^3} - \frac{X_1 - X_4}{R_{14}^3}\right) \tag{18.7}$$

$$\frac{d^2 Y_1}{dT^2} = (2.5324576)\left(-\frac{Y_1 - Y_2}{R_{12}^3} + \frac{Y_1 - Y_3}{R_{13}^3} - \frac{Y_1 - Y_4}{R_{14}^3}\right) \tag{18.8}$$

$$\frac{d^2 Z_1}{dT^2} = (2.5324576)\left(-\frac{Z_1 - Z_2}{R_{12}^3} + \frac{Z_1 - Z_3}{R_{13}^3} - \frac{Z_1 - Z_4}{R_{14}^3}\right) \tag{18.9}$$

$$\frac{d^2 X_2}{dT^2} = (1.379269)10^{-3}\left(-\frac{X_2 - X_1}{R_{12}^3} - \frac{X_2 - X_3}{R_{23}^3} + \frac{X_2 - X_4}{R_{24}^3}\right) \tag{18.10}$$

$$\frac{d^2 Y_2}{dT^2} = (1.379269)10^{-3}\left(-\frac{Y_2 - Y_1}{R_{12}^3} - \frac{Y_2 - Y_3}{R_{23}^3} + \frac{Y_2 - Y_4}{R_{24}^3}\right) \tag{18.11}$$

$$\frac{d^2 Z_2}{dT^2} = (1.379269)10^{-3}\left(-\frac{Z_2 - Z_1}{R_{12}^3} - \frac{Z_2 - Z_3}{R_{23}^3} + \frac{Z_2 - Z_4}{R_{24}^3}\right) \tag{18.12}$$

$$\frac{d^2 X_3}{dT^2} = 2.5324576\left(\frac{X_3 - X_1}{R_{13}^3} - \frac{X_3 - X_2}{R_{23}^3} - \frac{X_3 - X_4}{R_{34}^3}\right) \tag{18.13}$$

$$\frac{d^2 Y_3}{dT^2} = 2.5324576\left(\frac{Y_3 - Y_1}{R_{13}^3} - \frac{Y_3 - Y_2}{R_{23}^3} - \frac{Y_3 - Y_4}{R_{34}^3}\right) \tag{18.14}$$

$$\frac{d^2 Z_3}{dT^2} = 2.5324576 \left(\frac{Z_3 - Z_1}{R_{13}^3} - \frac{Z_3 - Z_2}{R_{23}^3} - \frac{Z_3 - Z_4}{R_{34}^3} \right) \tag{18.15}$$

$$\frac{d^2 X_4}{dT^2} = (1.379269)10^{-3} \left(-\frac{X_4 - X_1}{R_{14}^3} + \frac{X_4 - X_2}{R_{24}^3} - \frac{X_4 - X_3}{R_{34}^3} \right) \tag{18.16}$$

$$\frac{d^2 Y_4}{dT^2} = (1.379269)10^{-3} \left(-\frac{Y_4 - Y_1}{R_{14}^3} + \frac{Y_4 - Y_2}{R_{24}^3} - \frac{Y_4 - Y_3}{R_{34}^3} \right) \tag{18.17}$$

$$\frac{d^2 Z_4}{dT^2} = (1.379269)10^{-3} \left(-\frac{Z_4 - Z_1}{R_{14}^3} + \frac{Z_4 - Z_2}{R_{24}^3} - \frac{Z_4 - Z_3}{R_{34}^3} \right) \tag{18.18}$$

It is system (18.7) and (18.18) which will be solved numerically by the implicit, conservative methodology of Sect. 2.3 from given initial data. For convenience, we set $\vec{V}_i = \left(\frac{dX_i}{dT}, \frac{dY_i}{dT}, \frac{dZ_i}{dT} \right)$ and observe that $\vec{v}_i = 10^{10} \vec{V}_i$.

Note finally that the total energy E of the system is given by

$$E = \frac{1}{2} (9.1085)10^{-28}(v_1^2 + v_3^2) + \frac{1}{2} (16724)10^{-28}(v_2^2 + v_4^2)$$

$$+ (23.06689)10^{-20} \left(-\frac{1}{r_{12}} + \frac{1}{r_{13}} - \frac{1}{r_{14}} - \frac{1}{r_{23}} + \frac{1}{r_{24}} - \frac{1}{r_{34}} \right), \tag{18.19}$$

or equivalently, by

$$E = \frac{1}{2} (9.1085)10^{-8}(V_1^2 + V_3^2) + \frac{1}{2} (16724)10^{-8}(V_2^2 + V_4^2)$$

$$+ (23.06689)10^{-8} \left(-\frac{1}{R_{12}} + \frac{1}{R_{13}} - \frac{1}{R_{14}} - \frac{1}{R_{23}} + \frac{1}{R_{24}} - \frac{1}{R_{34}} \right)., \tag{18.20}$$

We consider now several examples. Assume that

$$\vec{R}_1 = (0, 6000, 0), \ \vec{R}_2 = (3742, 0, 0), \ \vec{R}_3 = -\vec{R}_1, \ \vec{R}_4 = -\vec{R}_2 \tag{18.21}$$

$$\vec{V}_1 = (0, 0, VZ), \ \vec{V}_2 = (VX, 0, 0), \ \vec{V}_3 = -\vec{V}_1, \ \vec{V}_4 = -\vec{V}_2 . \tag{18.22}$$

First, set $VX = 0.00025$. Since the system energy is $-(5.1104)10^{-11}$erg, substitution into (18.20) yields $VZ = 0.0143997$. Thus, all initial data are known. The system (18.7) and (18.18) was then solved numerically with $\Delta T = 2.0$, 1.0 and 0.5. We report only on the 0.5 case, which was the most accurate. The numerical solution was generated for 10^8 time steps. At each time step, the resulting nonlinear algebraic system was solved by Newtoian iteration with tolerances 10^{-8} for position and 10^{-11} for velocity. The average molecular diameter which resulted was 0.76 Å and the frequency of oscillation was $(2.1)10^{14} Hz$. Recall that the average diameter is 0.74 Å and frequency is $(1.3)10^{14} Hz$.

Though changes in VX and VZ did not alter the frequency by more than $(0.1)10^{14} Hz$, they did alter the molecular diameter more extensively. Thus, the choice $VX = 0.0003$, $VZ = 0.0125244$, which increased the initial speed of the protons yielded a frequency of $(2.1)10^{14} Hz$ and a molecular diameter of 0.82 Å.

On the other hand, the choice of $VX = 0.00015$, $VZ = 0.0167569$ yielded a frequency of $(2.2)10^{14} Hz$ and a molecular diameter of 0.66 Å.

In all cases, the errors in calculating the frequency were exorbitant.

18.3 Modification of the Classical Model

Since quantum mechanics implies that two electrons in the same orbital repel with an effective force which is less than that of full Coulombic repulsion, we repeated the classical calculation, but decreased the repulsive electron force by the factor 0.9. Assuming conservation of energy, we adjusted the initial velocities of the electrons accordingly. The vibrational frequency then decreased to $2.13 \times 10^{14} Hz$. Encouraged by this reduction, we proceeded in the above spirit to decrease electron repulsion until the factor 0.9 was reduced to 0.0001, but the vibrational frequency decreased only to $1.78 \times 10^{14} Hz$. We then proceeded through zero to choose negative factors until the Coulombic force between the electrons was multiplied by -1.0, that is, until the force between the electrons was assumed to be fully attractive rather than fully repulsive. The final results were correct. We proceed then to modify the discussion in Sect. 18.2 so that the electrons attract rather than repel.

It should be pointed out immediately that electron attraction is not unknown. For example, a quantum theory of superconductivity requires electron attraction (Bardeen, Cooper and Schrieffer (1957)).

The basic changes to be made then are as follows. In system (18.1), the term e_1e_3 has to be replaced by $-e_1e_3$. The formulas (18.19) and (18.20) are then replaced by

$$E = \frac{1}{2} \times 9.1085 \times 10^{-28}(v_1^2 + v_3^2) + \frac{1}{2}(16724) \times 10^{-28}(v_2^2 + v_4^2)$$

$$+ 23.06689 \times 10^{-20}\left(-\frac{1}{r_{12}} - \frac{1}{r_{13}} - \frac{1}{r_{14}} - \frac{1}{r_{23}} + \frac{1}{r_{24}} - \frac{1}{r_{34}}\right)$$

and

$$E = \frac{1}{2} \times 9.1085 \times 10^{-8}(V_1^2 + V_3^2) + \frac{1}{2} \times 16724 \times 10^{-8}(V_2^2 + V_4^2)$$

$$+ 23.06689 \times 10^{-8}\left(-\frac{1}{R_{12}} - \frac{1}{R_{13}} - \frac{1}{R_{14}} - \frac{1}{R_{23}} + \frac{1}{R_{24}} - \frac{1}{R_{34}}\right),$$

respectively.

Table 18.1 then records the resulting average vibrational frequencies f and diameters for the indicated parameters X, Y, VX, with $VX = 0$. The conservative method of Sect. 2.3 is, of course, essential, since the ground state energy is time invariant. The results all are entirely within physically acceptable scientific limits (Greenspan (1992)).

Table 18.1

Case	X	Y	VZ	$f(10^{14}Hz)$	$d(\text{Å})$
1	4000	4500	0.033 020 594	1.366	0.776
2	4000	4200	0.034 271 092	1.375	0.770
3	4000	4000	0.035 125 168	1.383	0.764
4	3800	5000	0.031 242 354	1.377	0.794
5	4435	5000	0.039 437 537	1.363	0.790
6	3000	4360	0.035 006 180	1.339	0.808
7	4000	10000	0.013 916 043	1.409	0.762

Recently (Greenspan (1997)), the method of this section has resulted in the correct calculation of the first excited state of H_2.

18.4 Extension to Li_2, B_2, C_2, N_2, and O_2

Classical calculation of correct frequencies and bond lengths for the diatomic molecules Li_2, B_2, C_2, N_2, and O_2 can be accomplished by the method of Sect. 18.3 if one proceeds as follows. For Li_2, B_2, C_2, and N_2, consider the nuclei and electrons arranged as shown in Fig. 18.1. The nuclei are denoted by P_1 and P_2. If in each case one allows attraction between pairs of electrons which are separated maximally, where one has $X < 0$ while the other has $X > 0$, then correct results follow (Greenspan (1993)). From Fig. 18.1 (a)–(d) one would guess that the use of hexagons would yield correct results for O_2. However, this is not the case. A more complex division of the electrons is required (Greenspan (1993b)).

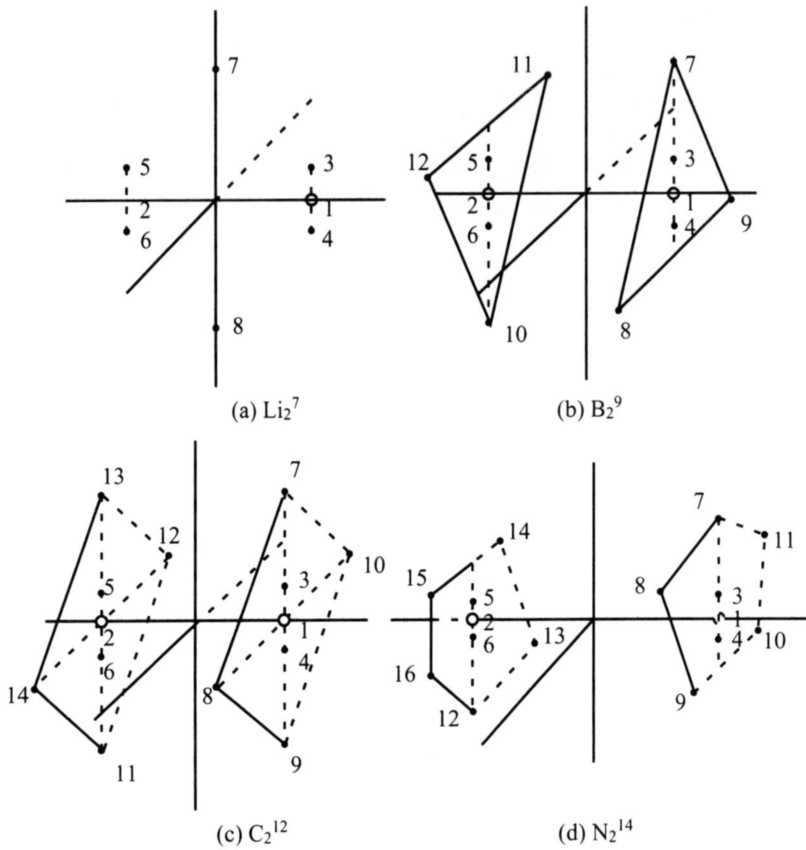

(a) Li_2^7

(b) B_2^9

(c) C_2^{12}

(d) N_2^{14}

Figure 18.1 — Electron and nuclei configurations for Li_2^7, B_2^9, C_2^{12}, N_2^{14}.

19

References and Sources for Further Reading

Adam, J. R., Lindblad, N. R., and Hendricks, C. D. (1968), The collision, coalescence, and disruption of water droplets, *J. Appl. Phys.* **39**, p. 5173.

Adam, N. K. (1930), *The Physics and Chemistry of Surfaces* (Clarendon Press, Oxford).

Adamson, A. W. (1976), *Physical Chemistry of Surfaces* (Interscience, NY).

Aguire-Puente, J., and Fremond, M. (1976), Frost propagation in wet porous media, in *Proceedings of Joint IUTAM/IMU Symposium on Application of Methods of Functional Analysis to Problems of Mechanics, Marseille, 1975* (Springer, Berlin).

Akiyoshi, T. (1978), Compatible viscous boundary for discrete models, *ASCE, J. Eng. Mech. Div.* p. 1253.

Albrycht, J. and Marciniak, A. (1981), Orbit calculations nearby the equilibrium points by a discrete mechanics method, *Cel. Mech.* **24**, p. 391.

Almgren, F. J. and Taylor, J. E. (1976), The geometry of soap films and soap bubbles, *Sci. Am.* **235**, p. 82.

Amsden, A. A. (1966), The particle-in-cell method for the calculation of the dynamics of compressible fluids, LA-3466, Los Alamos Sci. Lab., Los Alamos, NM.

Antonelli, P., Rogers, T. D., and Willard, M. (1973), Cell aggregation kinetics, *J. Theor. Biol.* **41**, p. 1.

Arzelies, H. (1966), *Relativistic Kinematics* (Pergamon, NY).

Ashurst, W. T., and Hoover, W. G. (1976), Microscopic fracture studies in the two-dimensional triangular lattice, *Phys. Rev.* **B14**, p. 1465.

Auret, F. D. and Snyman, J. A. (1978), Numerical study of linear and non-linear string vibrations by means of physical discretization, *Appl. Math. Modelling* **2**, p. 7.

Bardeen, J., Cooper, L. N., and Schrieffer, J. R. (1957), Theory of superconductivity, *Phys. Rev.* **108**, p. 1175.

Barenblatt, G. I., Looss, G., and Joseph, D. D. (1983), *Nonlinear Dynamics and Turbulence* (Pitman, Boston).

Barker, J. A., and Henderson, D. (1976), What is 'liquid'? Understanding the states of matter, *Rev. Mod. Phys.* **48**, p. 587.

Barto, A. G. (1975), Cellular Automata as Models of Natural Systems, Ph.D. thesis, Univ. Michigan, Ann Arbor.

Bergmann, P. G. (1942), *Introduction to the Theory of Relativity* (Prentice-Hall, Englewood Cliffs, NJ).

Birkhoff, G. and Lynch, R. E. (1961), Lagrangian hydrodynamic computations and molecular models of matter, LA-2618, Los Alamos Sci. Lab., Los Alamos, NM.

Bombolakis, E. G. (1968), Photoelastic study of initial stages of brittle fracture in compression, *Int. J. Tectonophys.* **6**, p. 461.

Bond, W. N. (1927), Bubbles and drops and Stokes' law, *Phil. Mag.* **4**, p. 889.

Boris, J. (1986), A vectorized 'near-neighbor' algorithm of order N using a monotonic logical grid, *J. Comp. Phys.* **66**, p. 1.

Borman, S. (1990), Theory, experiment team up to probe 'simplest' reaction, *Chem. and Eng. News* **4**, p. 32.

Boussinesq, J. (1913), Contribution to the theory of capillary action with an extension of viscous forces to the surface layers of liquids and an application notably to the slow uniform motion of a spherical fluid drop, *Comp. Rend.* **156**, p. 1124.

Broberg, K. B. (1971), Crack-growth criteria in non-linear fracture mechanics, *J. Mech. Phys. Solids* **19**, p. 407.

Bulgarelli, U., Casulli, V., and Greenspan, D. (1984), *Pressure Methods for the Numerical Solution of Free Surface Fluid Flows* (Pineridge, Swansea).

Buneman, O., Barnes, C. W., Green, J. C., and Nielsen, D. E. (1980), Principles and capabilities of 3-D, E-M particle simulations, *J. Comp. Phys.* **38**, p. 1.

Burridge, R. (ed.) (1978), *Fracture Mechanics*, (Am. Math. Soc., Providence, RI).

Cadzow, J. A. (1970), Discrete calculus of variations, *Int. J. Control* **11**, p. 393.

Cauchy, A. L. (1828), Sur l'equilibre et le mouvement d'un systeme de points materials sollicités par des forces d'attraction ou de repulsion mutuelles, *Exerc. de Math.* **3**.

Ciavaldini, J. F. (1975), Analyse numerique d'un problème de Stefan, *SIAM J. Numer. Anal.* **12**, p. 464.

Concus, P. (1967), Numerical solution of the minimal surface equation, *Math. Comp.* **21**, p. 340.

Coppin, C. and Greenspan, D. (1983), Discrete modelling of minimal surfaces, *Appl. Math. Comp.* **13**, p. 17.

Coppin, C. and Greenspan, D. (1988), A contribution to the particle modelling of soap films, *Appl. Math. Comp.* **26**, p. 315.

Costabel, P. (1973), *Leibniz and Dynamics* (Cornell University Press, NY).

Cotterell, B. (1972), Brittle fracture in compression, *Int. J. Fracture Mech.* **8**, p. 195.

Cotterill, R. M. J., Kristensen, W. D., and Jensen, E. J. (1974), Molecular dynamics studies of melting. III. Spontaneous dislocation generation and the dynamics of melting, *Phil. Mag.* **30**, p. 245.

Courant, R. (1950), *Dirichlet's Principle, Conformal Mapping and Minimal Surfaces* (Interscience, NY).

Crank, J. (1957), Two methods for the numerical solution of moving-boundary problems in diffusion and heat flow, *Quart. J. Mech. Appl. Math.* **10**, p. 220.

Cryer, C. W. (1969), Stability analysis in discrete mechanics, TR#67, Computer Sciences Department, University of Wisconsin, Madison, WI.

Daly, B. J. (1969), A technique for including surface tension effects in hydrodynamic calculations, *J. Comput. Phys.* **4**, p. 97.

Daly, B. J., Harlow, F. H., and Welch, J. E. (1965), Numerical fluid dynamics using the particle-and-force method, LA-3144, Part I, Los Alamos Sci. Lab., Los Alamos, NM.

Das, S. and Aki, K. (1977), Fault plane with barriers: A versatile earthquake model, *J. Geophys. Res.* **82**, p. 5658.

Davis, R. W. and Moore, E. F. (1982), A numerical study of vortex shedding from rectangles, *J. Fluid Mech.* **116**, p. 475.

de Celis, B., Argon, A. S., and Yip, S. (1983), Molecular dynamics simulation of crack tip processes in alpha-iron and copper, *J. Appl. Phys.* **54**, p. 48.

Dean, J. A. (ed.) (1985), *Lange's Handbook of Chemistry*, 13th Edition (McGraw-Hill, New York).

Denavit, J. (1974), Discrete particle effects in whistler simulation, *J. Comp. Phys.* **15**, p. 449.

Dezin, A. A. (1984), Discrete models in mathematical physics, in *Current Problems in Mathematical Physics and Numerical Mathematics* (Nauka, Moscow), p. 75.

Dickey, R. W. (1966), Dynamic behavior of soap films, *Quart. J. Appl. Math.* **24**, p. 97.

Dienes, G. J. and Paskin, A. (1983), Computer modelling of cracks, in *Atomistics of Fracture* (Plenum, New York), p. 671.

Dryden, H. L., Murnagham, F. D., and Bateman, H. (1956), *Hydrodynamics* (Dover, New York).

Dukawicz, J. K. (1980), A particle-fluid numerical model for liquid sprays, *J. Comp. Phys.* **35**, p. 229.

Erickson, R. O. (1973), Tabular packing of spheres in biological fine structure, *Science* **181**, p. 705.

Eringen, A. C. (1968), Mechanics of micromorphic continua; mechanics of generalized continua, in *Proceedings IUTAM Symposium* (Springer, Berlin), p. 18.

Evans, D. J. and Hoover, W. G. (1986), Flows far from equilibrium via molecular dynamics, *Ann. Rev. Fluid Mech.* **18**, p. 243.

Favre, A. (ed.) (1964), *The Mechanics of Turbulence* (Gordon and Breach, New York).

Fermi, E., Pasta, J. and Ulam, S. (1955), Studies of nonlinear problem-I, LA-1940, Los Alamos Sci. Lab., Los Alamos, NM.

Feynman, R. P., Leighton, R. B., and Sands, M. (1963), *The Feynman Lectures on Physics* (Addison-Wesley, Reading, MA).

Finn, R. (1986), *Equilibrium Capillary Surfaces* (Springer-Verlag, New York).

Freund, L. B. (1978), Stress intensity factor calculations based on a conservative integral, *Int. J. Solids Struct.* **14**, p. 241.

Froda, A. (1977), La finitude en mecanique classique, ses axiomes es leurs implications, Symposium on the Axiomatic Method, p. 238.

Gell-Mann, M. and Hartle, J. B. (1993), Classical equations for quantum systems, Phys. Rev. D *47* (1993) p. 3345.

Gentil, S. (1979), A discrete model for the study of a lake, *Appl. Math. Modelling* **3**, p. 193.

Girifalco, L. A. and Lad, R. A. (1956), Energy of cohesion, compressibility, and the potential energy functions of the graphite system, *J. Chem. Phys.* **25**, p. 693.

Goel, N. S., Campbell, R. D., Gordon, R. D., Rosen, R., Martinez, H., and Ycas, M. (1970), Self-sorting of isotropic cells, *J. Theor. Biol.* **28**, p. 423.

Goel, N. S. and Leith, A. G. (1970), Self-sorting of anisotropic cells, *J. Theor. Biol.* **28**, p. 469.

Goel, N. S. and Rogers, G. (1978), Computer simulation of engulfment and other movements of embryonic tissues, *J. Theor. Biol.* **71**, p. 103.

Goldstein, H. (1980), *Classical Mechanics*, 2nd Edition (Addison-Wesley, Reading, MA).

Gordon, R., Goel, N. S., Steinberg, M. S., and Wiseman, L. L. (1972), A rheological mechanism sufficient to explain the kinetics of cell sorting, *J. Theor. Biol.* **37**, p. 43.

Gottlieb, M. (1977), Application of computer simulation techniques to macro-molecular theories, *Comput. Chem.* **1**, p. 155.

Gotusso, L. and Veneziani, A. (1994), Discrete and continuous nonlinear models for the vibrating string, n. 143/P, Dip. Mat., Politecnico di Milano, P. L. da V., 32-20133 Milano, Italy.

Gray, A. (1959), *Gyrostatics and Rotational Motion* (Dover, New York).

Greenspan, D. (1967), On approximating extremals of functionals, II, *Int. J. Engrg. Sci.* **5**, p. 571.

Greenspan, D. (1968a), *Introduction to Calculus* (Harper and Row, New York).

Greenspan, D. (1968b), Discrete mechanics, Tech. Rpt. 49, Dept. Comp. Sci., Univ. Wisconsin, Madison.

Greenspan, D. (1970), Discrete, nonlinear string vibrations, *The Computer J.* **13**, p. 195.

Greenspan, D. (1971a), Computer simulation of transverse string vibrations, *BIT* **11**, p. 399.

Greenspan, D. (1971b), Computer power and its impact on applied mathematics, in *Studies in Mathematics Vol. 7*, A. H. Taub, ed. (Prentice-Hall, Englewood Cliffs, NJ), p. 65.

Greenspan, D. (1972a), Discrete Newtonian gravitation and the N-body problem, *Utilitas Math.* **2**, p. 105.

Greenspan, D. (1972b), An energy conserving, stable discretization of the harmonic oscillator, *Bull. Poly. Inst. Iasi,* XVII(XXII), Section 1, p. 205.

Greenspan, D. (1972c), Discrete solitary waves, TR167, Dept. Comp. Sci., University of Wisconsin, Madison.

Greenspan, D. (1972d), A discrete numerical approach to fluid dynamics, in *Proceedings IFIPS 71* (North-Holland).

Greenspan, D. (1972e), New forms of discrete mechanics, *Kybernetes* **1**, p. 87.

Greenspan, D. (1972f), A new explicit discrete mechanics with applications, *J. Franklin Inst.* **294**, p. 231.

Greenspan, D. (1973a), *Discrete Models* (Addison-Wesley, Reading, MA).

Greenspan, D. (1973b), Symmetry in discrete mechanics, *Found. Phys.* **3**, p. 247.

Greenspan, D. (1973c), A finite difference proof that $E = mc^2$, *Am. Math. Mon.* **80**, p. 289.

Greenspan, D. (1973d), An algebraic, energy conserving formulation of classical molecular and Newtonian N-body interaction, *Bull. AMS* **79**, p. 423.

Greenspan, D. (1974a), Discrete Newtonian gravitation and the three-body problem, *Found. Phys.* **4**, p. 299.

Greenspan, D. (1974b), Discrete bars, conductive heat transfer, and elasticity, *Comp. Struct.* **4**, p. 243.

Greenspan, D. (1974c), A physically consistent, discrete N-body model, *Bull. Am. Math. Soc.* **80**, p. 553.

Greenspan, D. (1974d), An arithmetic, particle theory of fluid dynamics, *Comp. Math. Appl. Mech. Eng.* **3**, p. 293.

Greenspan, D. (1974e), *Discrete Numerical Methods in Physics and Engineering* (Academic Press, New York).

Greenspan, D. (1975), Computer Newtonian and special relativistic mechanics, in *Proceedings of the Second USA-Japan Computer Conference*, (Am. Fed. Inf. Proc. Soc., Montvale, NJ), p. 88.

Greenspan, D. (1976), The arithmetic basis of special relativity, *Int. J. Theor. Phys.* **15**, p. 557.

Greenspan, D. (1977a), Conservative discrete models with computer examples of nonlinear phenomena in solids and fluids, *Am. J. Phys.* **45**, p. 740.

Greenspan, D. (1977b), Computer studies of interaction of particles with differing masses, *J. Comp. Appl. Math.* **3**, p. 145.

Greenspan, D. (1977c), On the arithmetic basis of special relativity, *Bul. Inst. Polit. Din. IASI*, XXIII (XXVII).

Greenspan, D. (1977d), Arithmetic applied mathematics, *Comp. Math. Appl.* **3**, p. 253.

Greenspan, D. (1977e), Computer studies of interactions of particles with differing masses, *J. Comp. Appl. Math.* **3**, p. 145.

Greenspan, D. (1978a), Computer studies of a von Neuman type fluid, *Appl. Math. Comp.* **4**, p. 15.

Greenspan, D. (1978b), A completely arithmetic formulation of classical and special relativistic mechanics, *Int. J. Gen. Syst.* **4**, p. 105.

Greenspan, D. (1978c), A particle model of the Stefan problem, *Comp. Meth. Appl. Mech. Eng.* **13**, p. 95.

Greenspan, D. (1980a), *Arithmetic Applied Mathematics* (Pergamon, Oxford).

Greenspan, D. (1980b), N-body modelling of nonlinear, free surface liquid flow, *Math. Comp. Simulation* **XXII**, p. 200.

Greenspan, D. (1980c), New mathematical models of porous flow, *Appl. Math. Modelling* **4**, p. 95.

Greenspan, D. (1980d), Discrete modelling in the microcosm and in the macrocosm, *Int. J. Gen. Syst.* **6**, p. 25.

Greenspan, D. (1981a), *Computer-Oriented Mathematical Physics* (Pergamon, Oxford).

Greenspan, D. (1981b), A classical molecular approach to computer simulation of biological sorting, *J. Math. Biol.* **12**, p. 227.

Greenspan, D. (1982a), Qualitative and quantitative particle modelling with applications to wave generation, vibration, and biomathematics, in *Dynamical Systems II* (Academic Press, NY), p. 71.

Greenspan, D. (1982b), Deterministic computer physics, *Int. J. Theor. Phys.* **21**, p. 505.

Greenspan, D. (1982c), Direct computer simulation of nonlinear waves in solids, liquids and gases, in *Nonlinear Phenomena in Mathematical Sciences* (Academic Press, NY), p. 471.

Greenspan, D. (1982d), Computer modelling of double-layer circularization and gastrulation, in *Discrete Simulation and Related Fields* (North-Holland), p. 153.

Greenspan, D. (1983a), A new computer approach to the modelling of close binary stars, *Astrophys. Space Sci.* **93**, p. 351.

Greenspan, D. (1983b), Computer-oriented, N-body modelling of minimal surfaces, *Appl. Math. Modelling* **7**, p. 423.

Greenspan, D. (1983c), Direct computer modelling, in *Mathematical Modelling in Science and Technology* (Pergamon, NY), p. 46.

Greenspan, D. (1983d), An arithmetic theory of gravity, in *Discrete and System Models* (Springer, NY), p. 46.

Greenspan, D. (1984a), Computer studies in particle modelling of fluid phenomena, *Math. Modelling* **6**, p. 273.

Greenspan, D. (1984b), A new mathematical approach to biological cell rearrangement with application to the inversion of volvox, *Syst. Anal. Model. Simul.* **1**, p. 5.

Greenspan, D. (1984c), Conservative numerical models for $\ddot{x} = f(x)$, *J. Comp. Phys.* **56**, p. 28.

Greenspan, D. (1985), Discrete mathematical physics and particle modelling, in *Adv. in Electronics and Electron Phys.* (Academic Press, NY), p. 189.

Greenspan, D. (1986a), Quasimolecular particle modelling of crack generation and fracture, *Comp. Struct.* **22**, p. 1055.

Greenspan, D. (1986b), Particle simulation of compression waves, *Math. Comp. Simul.* **28**, p. 13.

Greenspan, D. (1987a), Discrete arithmetic based simulation modelling formalism, in *Encyclopedia of Systems and Control* (Pergamon, Oxford), p. 4345.

Greenspan, D. (1987b), Quasimolecular modelling of turbulent and nonturbulent vortices, *Appl. Math. Modelling* **11**, p. 465.

Greenspan, D. (1987c), Particle modelling by systems of nonlinear ordinary differential equations, in *Nonlinear Analysis and Applications* (Dekker, NY), p. 203.

Greenspan, D. (1987d), Particle simulation of spiral galaxy evolution, *Math. Modelling* **9**, p. 785.

Greenspan, D. (1988a), Quasimolecular channel and vortex street modelling on a supercomputer, *Comp. Math. Appl.* **15**, p. 141.

Greenspan, D. (1988b), Mechanisms of capillarity via supercomputer simulation, *Comp. Math. Appl.* **16**, p. 331.

Greenspan, D. (1988c), Supercomputer simulation of sessile and pendent drops, *Math. Comp. Modelling* **10**, p. 871.

Greenspan, D. (1988d), Quasimolecular modelling of the cavity problem on a vector computer, *Appl. Math Modelling* **12**, p. 305.

Greenspan, D. (1988e), Particle modelling of cavity flow on a vector computer, *Comp. Meths. Appl. Math. Eng.* **66**, p. 291.

Greenspan, D. (1988f), Particle simulation of biological sorting on a supercomputer, TR#254, Math. Dept., Univ. Texas at Arlington.

Greenspan, D. (1989a), Quasimolecular simulation of large liquid drops, *J. Phys. D.: Appl. Phys.* **22**, p. 1415.

Greenspan, D. (1989b), Supercomputer simulation of cracks and fractures by quasimolecular dynamics, *J. Phys. Chem. Solid* **50**, p. 1245.

Greenspan, D. (1990), Supercomputer simulation of colliding microdrops of water, *Comp. Math. Appl.* **19**, p. 91.

Greenspan, D. (1992), Electron attraction as a mechanism for the molecular bond, *Phys. Essays* **5**, p. 250.

Greenspan, D. (1993), Electron attraction and Newtonian methodology for approximating quantum mechanical phenomena, *Comp. Math. Appl.* **25**, p. 75.

Greenspan, D. (1995), Completely conservative, covariant numerical methodology, *Comp. Math. Appl.* **29**, p. 37.

Greenspan, D. (1997), Dynamical generation of electron motions in ground state H_2^+, in ground state H_2, and in the first excited state of H_2, Physica Scripta, **55**, p.277.

Greenspan, D. and Casulli, V. (1985), Particle modelling of an elastic arch, *Appl. Math. Modelling* **9**, p. 215.

Greenspan, D. and Collier, J. (1978a), Computer studies of swirling particle fluids and the evolution of planetary-type bodies, *JIMA* **22**, p. 235.

Greenspan, D. and Collier, J. (1978b), Computer studies of planetary-type evolution, *J. Comp. Appl. Math.* **4**, p. 235.

Greenspan, D., Cranmer, M. and Collier, J. (1976), A particle model of ocean waves generated by earthquakes, Tech. Rpt. 277, Dept. Comp. Sci., Univ. Wisconsin, Madison.

Greenspan, D. and Heath, L. F. (1991), Supercomputer simulation of the modes of colliding microdrops of water, *J. Phys: D* **24**, p. 2121.

Greenspan, D. and Hougum, C. (1978), New investigations of von Neumann type fluids, TR 323, Dept. Comp. Sci., Univ. Wisconsin, Madison.

Greenspan, D. and Rosati, M. (1978), Computer generation of particle solids, *Comp. and Struct.* **8**, p. 107.

Ha, S. N. (1990), Experimental numerical studies on a supercomputer of natural convection in an enclosure with localized heating, *Comp. Math. Appl.* **20**, p. 1.

Haberman, R. (1987), *Elementary Applied Partial Differential Equations*, 2nd Edition (Prentice Hall, Englewood Cliffs, NJ).

Halicioglu, T. and Cooper, D. M. (1984), An atomistic model of slip formation, *Mater. Sci. Eng.* **62**, p. 121.

Harlow, F. H. and Sanmann, E. E. (1965), Numerical fluid dynamics using the particle-and-force method, LA-3144, Los Alamos Sci. Lab., Los Alamos, NM.

Harlow, F. H. and Shannon, J. P. (1967), The splash of a liquid drop, *J. Appl. Phys.* **38**, p. 3855.

Hertzberg, R. W. (1976), *Deformation and Fracture Mechanics of Engineering Materials* (Wiley, NY).

Herzberg, G. (1965), *Molecular Spectra and Molecular Structure*, 2nd Edition (van Nostrand, NY).

Hess, S. (1985), Dynamics of dense systems of spherical particles under shear, *J. Mec. Theor. et Appliq.* Special Number, p. 1.

Hida, K. and Nakanishi, T. (1970), The shape of a bubble or a drop attached to a flat plate, *J. Phys. Soc.* **28**, Japan, p. 1336.

Hilbert, D. and Cohn-Vossen, S. (1956), *Geometry and the Imagination* (Chelsea, NY).

Hinata, M., Shimasaki, M. and Kiyono, T. (1974), Numerical solution of the Plateau problem by a finite element method, *Math. Comp.* **28**, p. 45.

Hirschfelder, J. O., Curtiss, C. F. and Bird, R. B. (1954), *Molecular Theory of Gases and Liquids* (Wiley, NY).

Hockney, R. W., and Eastwood, J. W. (1981), *Computer Simulation Using Particles* (McGraw-Hill, NY).

Holtfreter, J. (1943), A study of the mechanics of gastrulation, *J. Exp. Biol.* **94**, p. 261.

Hopf, E. (1948), A mathematical example displaying features of turbulence, *Comm. Pure Appl. Math.* **1**, p. 303.

Huang, K. (1950), On the atomic theory of elasticity, *Proc. Roy. Soc. London* **A203**, p. 178.

Hudson, J. A., Hardy, M. P. and Fairhurst, C. (1973), The failure of rock beams: Part I-Theoretical studies, *Int. J. Rock Mech. Min. Sci.* **10**, p. 69.

Ikeda, T. (1986), A discrete model for spatially aggregating phenomena, in *Studies Math. Appl.* **18** (North-Holland), p. 385.

Jackson, J. C. (1977), A quantization of time, *J. Phys. A: Math. Gen.* **10**, p. 2115.

Kanatani, K. (1984), The accuracy and the preservation property of the discrete mechanics, *J. Comp. Phys.* **53**, p. 181.

Kanninen, M. F. (1978), A critical appraisal of solution techiques in dynamic fracture mechanics, in *Numerical Methods in Fracture Mechanics*, (Univ. Coll. of Swansea, Swansea, UK).

Kanury, A. M. (1975), *Introduction to Combustion Phenomena* (Gordon and Breach, NY).

Kardestuncer, H. (1975), *Discrete Mechanics—A Unified Approach* (Springer-Verlag, NY).

Keller, H. B. and Reiss, E. L. (1959), Spherical cap snapping, *J. Aero/Space Sci.* **26**, p. 643.

Kelly, A. and Macmillan, N. H. (1986), *Strong Solids*, 3rd Edition (Clarendon Press, Oxford).

Kelly, B. T. (1981), *Physics of Graphite* (Applied Sci., London).

Kelly, T. D. (1991), Particle modelling of an elastic arch in three dimensions, *Comp. Math. Appl.* **22**, p. 47.

Killand, J. L. (1964), *Inversion of Volvox* (Univ. Microfilms, Ann Arbor, MI).

Kittel, C. (1971), *Introduction to Solid State Physics*, 4th Edition (Wiley, NY).

Kobayashi, A. S., Wade, B. G. and Maiden, D. E. (1972), Photoelastic investigation on the crack-arrest capability of a hole, *Exp. Mech.* **12**, p. 32.

Kopal, Z. (1978), *Dynamic of Close Binary Systems* (D. Reidel Publ. Co., Dordrecht, Holland).

Korlie, M., (1966), Particle Modeling of a Liquid Drop Formation on a Solid Surface in 3-D, Ph.D. thesis, Math., UT Arlington.

LaBudde, R. A. (1980), Discrete Hamiltonian mechanics, *Int. J. Gen. Syst.* **6**, p. 3.

LaBudde, R. A. and Greenspan, D. (1974a), Discrete mechanics for nonseparable potentials with application for the LEPS form, TR 210, Dept. Comp. Sci., Univ. Wisconsin, Madison.

LaBudde, R. A. and Greenspan, D. (1974b), Discrete mechanics—a general treatment, *J. Comp. Phys.* **15**, p. 134.

LaBudde, R. A. and Greenspan, D. (1976a), Energy and momentum conserving methods of arbitrary order for the numerical integration of equations of motion-I, *Numerische Math.* **25**, p. 323.

LaBudde, R. A. and Greenspan, D. (1976b), Energy and momentum conserving methods of arbitrary order for the numerical integration of equations of motion-II, *Numerische Math.* **26**, p. 1.

LaBudde, R. A. and Greenspan, D. (1978), Discrete mechanics for anisotropic potentials, *Virginia J. Sci.* **29**, p. 18.

LaBudde, R. A. and Greenspan, D. (1987), An energy conserving modification of numerical methods for the integration of equations of motion, *Int. J. Math. Math. Sci.* **10**, p. 173.

Landau, L. D. (1944), On the problem of turbulence, *Dokl. Akad. Nauk USSR* **44**, p. 311.

Landau, L. D. and Lifshitz, E. M. (1976), *Mechanics*, 3rd Edition (Pergamon, Oxford), p. 112.

Langdon, A. B. (1973), 'Energy-Conserving' plasma simulation algorithms, *J. Comp. Phys.* **12**, p. 247.

Lax, M. (1965), The relation between microscopic and macroscopic theories of elasticity, in *Lattice Dynamics* (Pergamon, Oxford), p. 583.

Leith, A. G. and Goel, N. S. (1971), Simulation of movement of cells during self sorting, *J. Theor. Biol.* **33**, p. 171.

Leontovich, A. M., Pyatetskii-Shapiro, I. I., and Stavskaya, O. N. (1971), The problem of circularization in mathematical modelling of morphogenesis, *Avtomatika i Telemekhanika* **2**, p. 100.

Liebowitz, H. (ed.) (1968), *Fracture, an Advanced Treatise* (Academic, NY).

Lorente, M. (1974), Bases for a discrete special relativity, Publ. #437, Center for Theoretical Physics, MIT, Cambridge, MA.

MacPherson, A. K. (1971), The formulation of shock waves in a dense gas using a molecular dynamics type technique, *J. Fluid Mech.* **45**, p. 601.

Madariago, R. (1976), Dynamics of an expanding circular fault, *Bull. Seism. Soc. Am.* **65**, p. 163.

Maeda, S. (1979), On quadratic invariants in a discrete model of mechanical systems, *Math. Japonica* **23**, p. 587.

Mahar, T. J. (1982a), Discrete conservative oscillators: periodic and asymptotically periodic solutions, *SIAM J. Numer. Anal.* **19**.

Mahar, T. J. (1982b), Discrete almost-linear oscillators, *SIAM J. Numer. Anal.* **19**, p. 237.

Mahmoudi, M. (1990), Particle modelling of fluid phenomena in 3-D, *Comp. Math. Appl.* **20**, p. 25.

Malkus, W. V. R. (1960), Summer Program Notes, Woods Hole Ocean. Inst., Woods Hole, MA.

Malone, G. H., Hutchinson, T. E. and Prager, S. (1974), Molecular models for permeation through thin membranes: the effect of hydrodynamic interaction on permeability, *J. Fluid Mech.* **65**, p. 753.

Marciniak, A. (1985), *Numerical Solutions of the N-Body Problem* (D. Reidel, Dordrecht).

Marciniak, A. and Greenspan, D. (1991), Arbitrary order, Hamiltonian conserving numerical solutions of Caologero and Toda systems, *Comp. Math. Appl.* **22**, p. 11.

Markatos, N. C. (1986), The mathematical modelling of turbulent flows, *Appl. Math. Modelling* **10**, p. 190.

Matela, R. J. and Fletterick, R. J. (1980), Computer simulation of cellular self-sorting, *J. Theor. Biol.* **84**, p. 673.

May, R. M. (1975), Biological populations obeying difference equations: stable points, stable cycles and chaos, *J. Theor. Biol.* **51**, p. 511.

Meyer, G. H. (1973), Multidimensional Stefan problems, *SIAM J. Numer. Anal.* **10**, p. 522.

Meyer, R. (1971), *Introduction to Mathematical Fluid Dynamics* (Wiley, NY).

Miller, R. H. and Alton, N. (1968), Three dimensional n-body calculations, *ICR Quart. Rpt.* **#18**, Univ. Chicago.

Mostow, G. D. (ed.) (1975), *Mathematical Models for Cell Rearrangement* (Yale Univ. Press, New Haven, CT).

Muetterties, E. L. (1977), Molecular metal clusters, *Science* **196**, p. 839.

Murdoch, A. I. (1985), A corpuscular approach to continuum mechanics: Basic considerations, *Arch. Rat. Mech. Anal.* **88**, p. 291.

Neumann, C. P., and Tourassis, V. D. (1985), Discrete dynamics robot models, *IEEE Trans. on Systems, Man and Cybernetics*, Vol. SMC 15, #2, March/April, p. 193.

Nichols, B. D. and Hirt, C. W. (1971), Improved free surface boundary conditions for numerical incompressible flow calculations, *J. Comp. Phys.* **8**, p. 434.

Okubo, A. (1980), *Diffusion and Ecological Problems: Mathematical Models* (Springer-Verlag, Berlin).

Pan, F. and Acrivos, A. (1967), Steady flows in rectangular cavities, *J. Fluid Mech.* **28**, p. 643.

Pasta, J. R. and Ulam, S. (1957), Heuristic numerical work in some problems of hydrodynamics, *MTAC* **13**, p. 1.

Pavlovic, M. N. (1986), A simple model for thin shell theory-Part 2: Discretized surface, bending theory, and membrane hypothesis, *Int. J. Mech. Eng. Ed.* **13**, p. 199.

Perrone, N. and Alturi, S. N. (eds.) (1979), *Nonlinear and Dynamic Fracture Mechanics* (ASME, NY).

Petersen, R. A. and Uccellini, L. (1979). The computation of isentropic atmospheric trajectories using a discrete model, *Monthly Weather Rev.* **107**, p. 566.

Petersen, I. (1985), Raindrop oscillation, *Sci. News* **2**, p. 136.

Phan-Thien, N. and Karihalov, B. L. (1982), Effective moduli of particulate solids, *ZAMM* **62**, p. 183.

Piest, J. (1974), Molecular fluid dynamics and theory of turbulent motion, *Physica* **73**, p. 474.

Polanyi, J. C. (1987), Some concepts in reaction dynamics, *Science* **236**, p. 680.

Pollard, H. (1976), *Celestial Mechanics*, Carus Monograph #18 (Math. Assoc. Am., Washington, DC).

Potter, D. (1973), *Computational Physics* (Wilcy, NY).

Prandtl, L. (1925), On the development of turbulence, *ZAMM* **5**, p. 136.

Preisendorfer, R. W. (1965), *Radiative Transfer in Discrete Spaces* (Pergamon, NY).

Rado, T. (1951), *On the Plateau Problem* (Chelsea, NY).

Raviart, P.-A. (1985), An analysis of particle methods, in *Lecture Notes in Mathematics* (Springer, NY), *Vol.***1127**, p. 243.

Rawlinson, J. S. and Widom, B. (1982), *Molecular Theory of Capillarity* (Clarendon Press, Oxford).

Reeves, W. R. and Greenspan, D. (1982), An analysis of stress wave propagation in slender bars using a discrete particle approach, *Appl. Math. Modelling* **6**, p. 185.

Roger, G. and Goel, N. S. (1978), Computer simulation of cellular movements: Cell sorting cellular migration through a mass of cells and contact inhibition, *J. Theor. Biol.* **71**, p. 141.

Rosenbaum, J. S. (1976), Conservation properties of numerical integration methods for systems of ordinary differential equations, *J. Comp. Phys.* **20**, p. 259.

Ruelle, D. and Takens, F. (1971), On the nature of turbulence, *Comm. Math. Phys.* **20**, p. 167.

Saffman, P. G. (1968), Lectures on homogeneous turbulence, in *Topics in Nonlinear Physics* (Springer-Verlag, NY), p. 485.

Sandlin, N. H. (1970), Master's thesis, Univ. Texas at Arlington.

Schlichting, H. (1960), *Boundary Layer Theory* (McGraw-Hill, NY).

Schubert, A. B. and Greenspan, D. (1972), Numerical studies of discrete vibrating strings, TR 158, Dept. Comp. Sci., Univ. Wisconsin, Madison.

Schwartz, H. M. (1968), *Introduction to Special Relativity* (McGraw-Hill, NY).

Sears, F. W. and Zemansky, M. W. (1957), *University Physics*, 2nd Edition (Addison-Wesley, Reading).

Shapiro, A. H. (ed.) (1972), *Illustrated Experiments in Fluid Mechanics* (MIT Press, Cambridge, MA).

Shibberu, Y. (1994), Discrete-Time Hamiltonian Dynamics, *Comp. Math. Appl.*, **28**, p. 123.

Sih, G. C. (ed.) (1973), *Dynamic Crack Propagation* (Noordhoff, Leiden).

Simpson, G. C. (1923), Water in the atmosphere, *Nature* **111**, p. 520.

Simpson, S. F. and Haller, F. J. (1988), Effects of experimental variables on mixing of solutions by collisions of microdroplets, *Analyt. Chem.* **60**, p. 2483.

Sneddon, I. N. and Lowengrub, M. (1969), *Crack Problems in the Classical Theory of Elasticity* (Wiley, NY).

Snyman, J. A. (1979), Continuous and discontinuous numerical solutions to the Troesch problem, *J. Comp. and Appl. Math.* **5**, p. 171.

Snyman, J. A. and Snyman, H. C. (1981), Computed epitaxial monolayer structures, *Surface Sci.* **105**, p. 357.

Snyman, J. A. and Vermeulen, P. J. (1979), Numerical determination of the configuration of heavy rotating chains, *Comp. Math. Appl.* **3**, p. 232.

Sokolnikoff, I. S. and Redheffer, R. M. (1966), *Mathematics of Physics and Modern Engineering*, 2nd Edition (McGraw-Hill, NY).

Soos, E. (1973), Discrete and continuous models of solids III, *Stud. Cerc. Mat.* **25**, p. 687.

Soules, T. F. and Bushey, R. F. (1983), The rheological properties and fracture of a molecular dynamic simulation of sodium silicate glass, *J. Chem. Phys.* **78**, p. 6307.

Stefan, J. (1889), Uber die Theorie der Eisbildung, insbesondere uber die Eisbildung im Polarmeere, Sitz Akad. Wiss. Wien. *Mat.-Nat. Classe* **98**, p. 965.

Steinberg, M. S. (1963), Reconstructing of tissues by dissociated cells, *Science* **141**, p. 401.

Su, C. H. and Mirie, R. M. (1980), On head-on collision between two solitary waves, *J. Fluid Mech.* **98**, p. 509.

Synge, J. L. (1965), *Relativity: The Special Theory* (North-Holland, Amsterdam).

Taylor, G. I. (1921), Diffusion by continuous movements, *Proc. London Math. Soc.* **A20**, p. 196.

Taylor, E. F. and Wheeler, J. A. (1966), *Spacetime Physics* (Freeman, San Francisco).

Teodorescu, P. P. and Soos, E. (1973), Discrete quasi-continuous and continuous models of elastic solids, *ZAMM* **53**, T33.

Toomre, A. and Toomre, J. (1973), Violent tides between galaxies, *Sci. Am.* **38**.

Trefethen, L. (1972), *Illustrated Experiments in Fluid Mechanics* (MIT Press, Cambridge, MA).

Uccellini, L. W. and Petersen, R. A. (1980), Applying discrete model concepts to the computation of atmospheric trajectories, *Int. J. Gen. Syst.* **6**, p. 13.

van Dyke, M. (1982), *An Album of Fluid Motion* (Parabolic Press, Stanford, CA).

Vargas, C. (1986), A discrete model for the recovery of oil from a reservoir, *Appl. Math. and Comp.* **18**, p. 93.

von Karman, T. (1963), *Aerodynamics* (McGraw-Hill, NY).

von Neumann, J. (1963), Proposal and analysis of a new numerical method for the treatment of hydrodynamical shock problems, in *The Collected Works of John von Neumann* Vol. 6, No. 27 (Pergamon, NY).

Vul, E. B. and Pyatitskii-Shapiro, I. I. (1971), A model of inversion in volvox, *Problemi Peredachi Informatsii* **7**, No. 4, p. 91.

Wadia, A. R. and Greenspan, D. (1980), An arithmetic approach to gas dynamical modelling, in *Innovative Numerical Analysis for the Engineering Sciences*, (Univ. Press of Virginia), p. 272.

Wagner, H.-J. (1977), A contribution to the numerical approximation of minimal surfaces, *Computing* **19**, p. 35.

Welch, J. E., Harlow, F. H., Shannon, J. P. and Daly, B. J. (1966), The MAC method, TR LA-3425, Los Alamos Sci. Lab., Los Alamos, NM.

Wente, H. C. (1980), The symmetry of sessile and pendent drops, *Pacific J. Math.* **88**, p. 387.

Whitrow, G. J. (1961), *The Natural Philosophy of Time* (Harper's, NY).

Winter, A. (1947), *The Analytical Foundations of Celestial Mechanics* (Princeton Univ. Press, Princeton, NJ).

Wyatt, B. M. (1994), Molecular Dynamics Simulation of Colliding Microdrops of Water, *Comp. Math. Appl.*, **28**, p. 175.

Young, D. M. and Gregory, R. T. (1972), *A Survey of Numerical Mathematics* (Addison-Wesley, Reading, MA).

Zeigler, B. P. (1976), *Theory of Modeling and Simulation* (Wiley, NY).

Zeldovich, Y. B., Barenblatt, G. I., Lebrovich, V. B. and Makhviladze, G. M. (1985), *The Mathematical Theory of Combustion and Explosions* (Consultants Bureau, NY).

Appendix A1
STRESS.FOR

```fortran
C  TOTAL NUMBER OF PARTICLES IS 226.
C P AND Q ARE MOLECULAR PARAMETERS 3 AND 5.
C THE CONE IS H20 MOLS IN 2D.
   DOUBLE PRECISION X0(226),Y0(226),VX0(226),VY0(226),
   1X(226,3),Y(226,3),VX(226,2),VY(226,2),
   1ACX(226),ACY(226),R2(226),R4(226),FX(226),FY(226),F(226)
   1,A(226,226),B(226,226)
   OPEN (UNIT=21,FILE='STRESS.DAT',STATUS='OLD')
   OPEN (UNIT=22,FILE='STRESS.OUT',STATUS='NEW')
   OPEN (UNIT=23,FILE='STRESS.XKE',STATUS='NEW')
   K=1
   DO 9900 I=1,226
   DO 9901 J=1,226
   IF (I.LE.7.AND.J.LE.7) A(I,J)=2500.6
   IF (I.GE.8.AND.J.GE.8) A(I,J)=2500.6
   IF (I.LE.7.AND.J.GE.8) A(I,J)=2500.6
   IF (I.GE.8.AND.J.LE.7) A(I.J)=2500.6
   IF (I.LE.7.AND.J.LE.7) B(I,J)=3541.2
   IF (I.GE.8.AND.J.GE.8) B(I,J)=3541.2
   IF (I.LE.7.AND.J.GE.8) B(I,J)=3541.2
   IF (I.GE.8.AND.J.LE.7) B(I,J)=3541.2
9901   CONTINUE
9900   CONTINUE
   READ (21,10) (X0(I),Y0(I),VX0(I),VY0(I),I=1,226)
10     FORMAT (4F16.10)
11     FORMAT (2F20.10)
265    FORMAT (I12,F30.5)
   DO 30 I=1,226
   X(I,1)=X0(I)
   Y(I,1)=Y0(I)
```

```fortran
      VX(I,1)=VX0(I)
      VY(I,1)=VY0(I)
30    CONTINUE
      GO TO 3456
65    DO 70 I=1,226
      X(I,1)=X(I,2)
      Y(I,1)=Y(I,2)
      VX(I,1)=VX(I,2)
      VY(I,1)=VY(I,2)
70    CONTINUE
      DO 701=1,226
      ACX(I)=0.
      ACY(I)=0.
      F(I)=0.
701     CONTINUE
3456    DO 78 I=1,225
      ACXI=ACX(I)
      ACYI=ACY(I)
      XI=X(I,1)
      YI=Y(I,1)
      IP1=I+1
      DO 77 J=IP1,226
      R2(J)=(XI-X(J,1))**2+(YI-Y(J,1))**2
C IN THIS PROGRAM THERE IS A LOCAL INTERACTION DISTANCE.
C THIS DISTANCE IS 1.6.
C HOWEVER WE USE SQUARES.
      IF (R2(J).GT.2.56) GO TO 77
      F(J)=(-A(I,J)+B(I,J)/(R2(J)))
    1    /(R2(J)*R2(J))
      ACX(J)=ACX(J)-F(J)*(XI-X(J,1))
      ACY(J)=ACY(J)-F(J)*(YI-Y(J,1))
      ACXI=ACXI+F(J)*(XI-X(J,1))
      ACYI=ACYI+F(J)*(YI-Y(J,1))
77  CONTINUE
      ACX(I)=ACXI
      ACY(I)=ACYI
78    CONTINUE
C INSERT GRAVITY.
      DO 1199 I=1,226
      ACY(I)=ACY(I)-32.667
1199  CONTINUE
      WRITE (22,11)(ACX(I),ACY(I).
     1I=1,226)
      STOP
      END
```

Appendix A2
DROP.FOR

```fortran
C  TOTAL NUMBER OF PARTICLES IS 4102.
C P AND Q ARE MOLECULAR PARAMETERS 3 AND 5.
      DIMENSION X0(4102),Y0(4102),VX0(4102),VY0(4102).
     1X(4102.3),Y(4102.3),VX(4102.2),VY(4102.2).
     1ACX(4102),ACY(4102),R2(4102),R4(4102),FX(4102),FY(4102),F(4102)
     1,Z0(4102),VZ0(4102),Z(4102.3),VZ(4102.2),ACZ(4102),FZ(4102)
      OPEN (UNIT=21)
      OPEN (UNIT=31)
      OPEN (UNIT=41)
      K=1
      KPRINT=1000
      DO 4 I=1,4102
      X(I,1)=0.
      Y(I,1)=0.
      Z(I,1)=0.
      VX(I,1)=0.
      VY(I,1)=0.
      VZ(I,1)=0.
      X(I,2)=0.
      Y(I,2)=0.
      Z(I,2)=0.
      VX(I,2)=0.
      VY(I,2)=0.
      VZ(I,2)=0.
      ACX(I)=0.
      ACY(I)=0.
      ACZ(I)=0.
4     CONTINUE
      READ (21,10) (X0(I),Y0(I),Z0(I),VX0(I),VY0(I),VZ0(I),I=1,4102)
10    FORMAT (6F16.10)
```

```fortran
      FORMAT (I12,F20.5)
      DO 30 I=1,4102
      X(I,1)=X0(I)
      Y(I,1)=Y0(I)
      Z(I,1)=Z0(I)
      VX(I,1)=VX0(I)
      VY(I,1)=VY0(I)
      VZ(I,1)=VZ0(I)
30    CONTINUE
      GO TO 3456
65    DO 70 I=1,4102
      X(I,1)=X(I,2)
      Y(I,1)=Y(I,2)
      Z(I,1)=Z(I,2)
      VX(I,1)=VX(I,2)
      VY(I,1)=VY(I,2)
      VZ(I,1)=VZ(I,2)
70    CONTINUE
      DO 701 I=1,4102
      ACX(I)=0.
      ACY(I)=0.
      ACZ(I)=0.0
      F(I)=0.
701   CONTINUE
3456    DO 78 I=1,4101
      ACXI=ACX(I)
      ACYI=ACY(I)
      ACZI=ACZ(I)
      XI=X(I,1)
      YI=Y(I,1)
      ZI=Z(I,1)
      IP1=I+1
      DO 77 J=IP1,4102
      R2(J)=(XI-X(J,1)**2+(YI-Y(J,1))**2
     1+(ZI-Z(J,1))**2
C IN THIS PROGRAM THERE IS NO LOCAL INTERACTION DISTANCE.
      F(J)=(-16.5+158.6/(R2(J)))
     1   /(R2(J)*R2(J))
      ACX(J)=ACX(J)-F(J)*(XI-X(J,1))
      ACY(J)=ACY(J)-F(J)*(YI-Y(J,1))
      ACZ(J)=ACZ(J)-F(J)*(ZI-Z(J,1))
      ACXI=ACXI+F(J)*(XI-X(J,1))
      ACYI=ACYI+F(J)*(YI-Y(J,1))
      ACZI=ACZI+F(J)*(ZI-Z(J,1))
77    CONTINUE
```

```fortran
      ACX(I)=ACXI
      ACY(I)=ACYI
      ACZ(I)=ACZI
78    CONTINUE
      DO 7123 I=1,4102
      VX(I,2)=VX(I,1)+0.0002*ACX(I)
      VY(I,2)=VY(I,1)+0.0002*ACY(I)
      VZ(I,2)=VZ(I,1)+0.0002*ACZ(I)
      X(I,2)=X(I,1)+0.0002*VX(I,2)
      Y(I,2)=Y(I,1)+0.0002*VY(I,2)
      Z(I,2)=Z(I,1)+0.0002*VZ(I,2)
7123  CONTINUE
      K=K+1
      IF (MOD(K,KPRINT).GT.0) GO TO 82
      WRITE (31,10) (X(I,2),Y(I,2),Z(I,2),VX(I,2),VY(I,2),
     1VZ(I,2),I=1,4102)
82    IF (K.LT.3001) GO TO 65
      STOP
      END
```

Appendix A3
MORSE.FOR

```fortran
      DOUBLE PRECISION X1,X2,X3,V1,V2,V3,XMIN,XMAX,X0.
     1V0,DELT,DIFFV,XX,VX,ENERGY
     1,C1,C2,C3,C4,C5,C6,C7,C8,C9,C10,C11,C12,C13
      OPEN (UNIT=21)
      OPEN (UNIT=31)
      OPEN (UNIT=41)
      C1=0.5
      C2=3.8919124
      C3=7.7838246
      C4=74.855925
C C5 AND C6 ARE CONVERGENCE TOLERANCES.
      C5=0.000000001
      C6=0.0000000001
      C7=0.16733
      C8=0.760429
      C9=8.4646357
      C10=4.2323178
      C11=4.35912
      C12=1.839145
      C13=17.912514
      KPRINT=100
      K=1
      XMIN=1000.
      XMAX=0.0
      DELT=.00001
C READ IN THE INITIAL DATA.
      READ (21,10) X0,V0
10    FORMAT(F16.11,F16.12)
C SET THE INPUT DATA.
      X1=X0
```

```fortran
      V1=V0
      GO TO 100
65    X1=X2
      V1=V2
C FIX THE FIRST GUESS X2 OF THE ITERATION.
100   X2=X1
      V2=V1
      KK=1
      GO TO 110
105   X2=X3
      V2=V3
110   X3=X1+C1*DELT*(V2+V1)
      DIFFV=(-DEXP(-C2*X3)+DEXP(-C2*X1))
     1/C2
      DIFFV=DIFFV-(-DEXP(-C3*X3)+DEXP(-C3*X1))
     1/C12
      DIFFV=C4*DIFFV
      DIFFV=DIFFV/(X3-X1)
      V3=V1-DELT*DIFFV
      KK=KK+1
      IF (KK.GT.1000) GO TO 9876
      XX=ABS(X3-X2)
      VX=ABS(V3-V2)
      INDEX=1
      IF (XX.GT.C5) INDEX=-1
      IF (VX.GT.C6) INDEX=-1
      IF (INDEX.EQ.-1) GO TO 105
      X2=X3
      V2=V3
      IF (X2.LT.XMIN) XMIN=X2
      IF (X2.GT.XMAX) XMAX=X2
9876  K=K+1
1999  IF (MOD(K,KPRINT).GT.0) GO TO 82
C CALCULATE AND PRINT FINAL ENERGY
C THE ENERGY IS 10**11ENERGY
      ENERGY=C7*(V2*V2)+C8*(-C9*
     1(DEXP(-C2*X2))+C13*(DEXP(-C3*X2)))
     1-C11
      WRITE (41,8112) X2,V2,K,ENERGY,XMIN,XMAX
8112    FORMAT (2F15.10,I10,F15.11,2F15.10)
8113  FORMAT (F25.15)
82    IF (K.LE.100000) GO TO 65
      WRITE (31,10) X2,V2
83    STOP
      END
```

Appendix A4
GHEXA.FOR

```fortran
C IN THIS PROGRAM WE DETERMINE INITIAL DATA FOR
C A SPINNING, REGULAR HEXAHEDRAL GYROSCOPE OF EDGE LENGTH
C R AND VERTICES P1 (X1,Y1,Z1)=(0,0,0), P2(X2,Y2,Z2),
C P3(X3,Y3,Z3), P4(X4,Y4,Z4), P5(X5,Y5,Z5)
      DOUBLE PRECISION XBAR,YBAR,ZBAR,R
     1,X1,Y1,Z1,X2,Y2,Z2,X3,Y3,Z3,X4,Y4,Z4
     1,C1,C2,C3,C4,C5,C6,C7,C8,C9,C10,C11,C12
     1,VX1,VX2,VX3,VX4,VY1,VY2,VY3,VY4,VZ1,VZ2,VZ3,VZ4
     1,ALPHA,XBARNEW,YBARNEW,ZBARNEW
     1,X1P,Y1P,Z1P,X2P,Y2P,Z2P,X3P,Y3P,Z3P,X4P,Y4P,Z4P,
     1VX1P,VY1P,VZ1P,VX2P,VY2P,VZ2P,VX3P,VY3P,VZ3P,
     1VX4P,VY4P,VZ4P
     1,X5,Y5,Z5,VX5,VY5,VZ5,X5P,Y5P,Z5P,VX5P,VY5P,VZ5P
      OPEN (UNIT=22,FILE='GHEXA.OUT',STATUS='NEW')
      OPEN (UNIT=23,FILE='GHEXA.XKE',STATUS='NEW')
      R=1.290994449
      V=40.0
      ALPHA=90.
      C1=3
      C1=DSQRT(C1)
      C2=6
      C2=DSQRT(C2)
      X1=0.0
      Y1=0.0
      Z1=0.0
      X2=0.0
      Y2=R*C1/3.
      Z2=R*C2/3.
      X3=.5*R
      Y3=-R*C1/6.
```

```
      Z3=R*C2/3.
      X4=-X3
      Y4=Y3
      Z4=Z3
      X5=0.
      Y5=0.
      Z5=2.*R*C2/3.
      VX1=0.
      VY1=0.
      VZ1=0.
      VX2=V
      VY2=0.
      VZ2=0.
      VX3=-.5*V
      VY3=-.5*C1*V
      VZ3=0.
      VX4=-.5*V
      VY4=.5*C1*V
      VZ4=0.
      VX5=0.
      VY5=0.
      VZ5=0.
      XBAR=(X1+X2+X3+X4+X5)/5.
      YBAR=(Y1+Y2+Y3+Y4+Y5)/5.
      ZBAR=(Z1+Z2+Z3+Z4+Z5)/5.
C WE NOW ROTATE THE X AND Z AXIS AN ANGLE ALPHA. THIS
C TILTS THE TETRAHEDRON. WE WILL GIVE ALPHA IN DEGREES
C BUT IT MUST BE TRANSFORMED INTO RADIANS.
      ALPHA=ALPHA*3.14159265358979/180.
C WE NOW TRANSFORM ALL POINTS AND VELOCITIES INTO
C THE PRIME COORDINATES.
      X1P=X1*COS(ALPHA)+Z1*SIN(ALPHA)
      Y1P=Y1
      Z1P=-X1*SIN(ALPHA)+Z1*COS(ALPHA)
      X2P=X2*COS(ALPHA)+Z2*SIN(ALPHA)
      Y2P=Y2
      Z2P=-X2*SIN(ALPHA)+Z2*COS(ALPHA)
      X3P=X3*COS(ALPHA)+Z3*SIN(ALPHA)
      Y3P=Y3
      Z3P=-X3*SIN(ALPHA)+Z3*COS(ALPHA)
      X4P=X4*COS(ALPHA)+Z4*SIN(ALPHA)
      Y4P=Y4
      Z4P=-X4*SIN(ALPHA)+Z4*COS(ALPHA)
      X5P=X5*COS(ALPHA)+Z5*SIN(ALPHA)
      Y5P=Y5
```

```
Z5P=-X5*SIN(ALPHA)+Z5*COS(ALPHA)
VX1P=VX1*COS(ALPHA)+VZ1*SIN(ALPHA)
VY1P=VY1
VZ1P=-VX1*SIN(ALPHA)+VZ1*COS(ALPHA)
VX2P=VX2*COS(ALPHA)+VZ2*SIN(ALPHA)
VY2P=VY2
VZ2P=-VX2*SIN(ALPHA)+VZ2*COS(ALPHA)
VX3P=VX3*COS(ALPHA)+VZ3*SIN(ALPHA)
VY3P=VY3
VZ3P=-VX3*SIN(ALPHA)+VZ3*COS(ALPHA)
VX4P=VX4*COS(ALPHA)+VZ4*SIN(ALPHA)
VY4P=VY4
VZ4P=-VX4*SIN(ALPHA)+VZ4*COS(ALPHA)
VX5P=VX5*COS(ALPHA)+VZ5*SIN(ALPHA)
VY5P=VY5
VZ5P=-VX5*SIN(ALPHA)+VX5*COS(ALPHA)
X1=X1P
Y1=Y1P
Z1=Z1P
X2=X2P
Y2=Y2P
Z2=Z2P
X3=X3P
Y3=Y3P
Z3=Z3P
X4=X4P
Y4=Y4P
Z4=Z4P
X5=X5P
Y5=Y5P
Z5=Z5P
XBARNEW=(X1+X2+X3+X4+X5)/5.
YBARNEW=(Y1+Y2+Y3+Y4+Y5)/5.
ZBARNEW=(Z1+Z2+Z3+Z4+Z5)/5.
VX1=VX1P
VY1=VY1P
VZ1=VZ1P
VX2=VX2P
VY2=VY2P
VZ2=VZ2P
VX3=VX3P
VY3=VY3P
VZ3=VZ3P
VX4=VX4P
VY4=VY4P
```

```
      VZ4=VZ4P
      VX5=VX5P
      VY5=VY5P
      VZ5=VZ5P
      WRITE (22,1001) X1,Y1,Z1,VX1,VY1,VZ1,
      WRITE (22,1001) X2,Y2,Z2,VX2,VY2,VZ2,
      WRITE (22,1001) X3,Y3,Z3,VX3,VY3,VZ3,
      WRITE (22,1001) X4,Y4,Z4,VX4,VY4,VZ4,
      WRITE (22,1001) X5,Y5,Z5,VX5,VY5,VZ5,
      WRITE (22,9999) XBAR,YBAR,ZBAR
      WRITE (22,9999) XBARNEW,YBARNEW,ZBARNEW
9999  FORMAT (3F15.10)
1001  FORMAT (3F16.10,3F16.10)
83    STOP
      END
```

HEXA.FOR

```
C IT IS IMPORTANT TO NOTE THAT THE IMPLICIT METHOD TO
C BE USED CONSERVES ENERGY, LINEAR MOMENTUM, AND ANGULAR
C MOMENTUM.
      DOUBLE PRECISION X0(5),Y0(5),VX0(5),VY0(5),
     1X(5,5),Y(5,5),VX(5,5),VY(5,5).
     1XX(5),YY(5),VXX(5),VYY(5)
     1,Z0(5),VZ0(5),Z(5,5),VZ(5,5),ZZ(5),VZZ(5)
     1,R(5,5),RN(5,5),RR(5,5),XXX(5,5),YYY(5,5),ZZZ(5,5)
     1,XTIME,DELT,ENERGY,V1SQ,V2SQ,V3SQ,V4SQ,V5SQ
     1,XBAR,YBAR,ZBAR
      OPEN (UNIT=21,FILE='HEXA.DAT',STATUS='OLD')
      OPEN (UNIT=22,FILE='HEXA.OUT',STATUS='NEW')
      OPEN (UNIT=23,FILE='HEXA.XKE',STATUS='NEW')
      OPEN (UNIT=24,FILE='HEXA.BAR',STATUS='NEW')
      A=1000000.
      KPRINT=10000
      K=1
      DELT=.00001
      XTIME=0.0
C READ IN THE INITIAL DATA.
      READ (21,10) (X0(I),Y0(I),Z0(I),VX0(I),VY0(I),VZ0(I),
     1I=1,5)
10    FORMAT (3F16.10,3F16.10)
C WE NOW SOLVE FOR THE NEW POSITIONS AND VELOCITIES WITH
```

```
C FORMULAS WHICH REQUIRE ITERATION. WE WILL USE THE
C IDEAS OF THE GENERALIZED NEWTON'S METHOD WITHOUT THE
C UPDATING EACH ITERATE.
C SET THE INPUT DATA.
      DO 30 I=1,5
      X(I,1)=X0(I)
      Y(I,1)=Y0(I)
      Z(I,1)=Z0(I)
      VX(I,1)=VX0(I)
      VY(I,1)=VY0(I)
      VZ(I,1)=VZ0(I)
30    CONTINUE
      GO TO 100
65    DO 70 I=1,5
      X(I,1)=X(I,2)
      Y(I,1)=Y(I,2)
      Z(I,1)=Z(I,2)
      VX(I,1)=VX(I,2)
      VY(I,1)=VY(I,2)
      VZ(I,1)=VZ(I,2)
70    CONTINUE
C FIX THE FIRST GUESS X(I,2) OF THE ITERATION.
100   DO 1633 I=1,5
      X(I,2)=X(I,1)
      Y(I,2)=Y(I,1)
      Z(I,2)=Z(I,1)
      VX(I,2)=VX(I,1)
      VY(I,2)=VY(I,1)
      VZ(I,2)=VZ(I,1)
1633  CONTINUE
      DO 103 I=1,5
      DO 102 J=1,5
      R(I,J)=(X(I,1)-X(J,1))**2+(Y(I,1)-Y(J,1))**2
     1+(Z(I,1)-Z(J,1))**2
      R(I,J)=DSQRT(R(I,J))
102   CONTINUE
103   CONTINUE
      XTIME=XTIME+DELT
      KK=1
      GO TO 110
105   DO 108 JJ=1,5
      X(JJ,2)=X(JJ,3)
      Y(JJ,2)=Y(JJ,3)
      Z(JJ,2)=Z(JJ,3)
      VX(JJ,2)=VX(JJ,3)
```

```fortran
      VY(JJ,2)=VY(JJ,3)
      VZ(JJ,2)=VZ(JJ,3)
108   CONTINUE
110   DO 203 I=1,5
      DO 202 J=1,5
      RN(I,J)=(X(I,2)-X(J,2))**2+(Y(I,2)-Y(J,2))**2
     1+(Z(I,2)-Z(J,2))**2
      RN(I,J)=DSQRT(RN(I,J))
202   CONTINUE
203   CONTINUE
      DO 109 I=1,5
      X(I,3)=X(I,1)+0.5*DELT*(VX(I,2)+VX(I,1))
      Y(I,3)=Y(I,1)+0.5*DELT*(VY(I,2)+VY(I,1))
      Z(I,3)=Z(I,1)+0.5*DELT*(VZ(I,2)+VZ(I,1))
109      CONTINUE
      DO 300 I=1,5
      DO 299 J=1,5
      XXX(I,J)=X(I,3)+X(I,1)-X(J,3)-X(J,1)
      YYY(I,J)=Y(I,3)+Y(I,1)-Y(J,3)-Y(J,1)
      ZZZ(I,J)=Z(I,3)+Z(I,1)-Z(J,3)-Z(J,1)
299   CONTINUE
300   CONTINUE
      DO 400 I=1,4
      DO 399 J=I+1,5
      RR(I,J)=-A*(R(I,J)**4+(R(I,J)**3)*RN(I,J)
     1+(R(I,J)**2)*(RN(I,J)**2)+R(I,J)*(RN(I,J)**3)
     1+(RN(I,J)**4)-(R(I,J)**2)*(RN(I,J)**2)*
     1(RN(I,J)**2+RN(I,J)*R(I,J)+R(I,J)**2))/
     1((RN(I,J)**5)*(R(I,J)**5)*(RN(I,J)+R(I,J)))
399   CONTINUE
400   CONTINUE
      VX(1,3)=VX(1,1)-DELT*(XXX(1,2)*RR(1,2) +XXX(1,3)*RR(1,3)
     1+XXX(1,4)*RR(1,4))
      VY(1,3)=VY(1,1)-DELT*(YYY(1,2)*RR(1,2) +YYY(1,3)*RR(1,3)
     1+YYY(1,4)*RR(1,4))
      VZ(1,3)=VZ(1,1)-DELT*(ZZZ(1,2)*RR(1,2) +ZZZ(1,3)*RR(1,3)
     1+ZZZ(1,4)*RR(1,4))
C NOTE THAT P1 IS INDEPENDENT OF GRAVITY.
      VX(2,3)=VX(2,1)-DELT*(XXX(2,1)*RR(1,2) +XXX(2,3)*RR(2,3)
     1+XXX(2,4)*RR(2,4)+XXX(2,5)*RR(2,5))
      VY(2,3)=VY(2,1)-DELT*(YYY(2,1)*RR(1,2) +YYY(2,3)*RR(2,3)
     1+YYY(2,4)*RR(2,4)+YYY(2,5)*RR(2,5))
      VZ(2,3)=VZ(2,1)-DELT*(ZZZ(2,1)*RR(1,2) +ZZZ(2,3)*RR(2,3)
     1+ZZZ(2,4)*RR(2,4)+ZZZ(2,5)*RR(2,5))
     1-DELT*.98
```

```
 VX(3,3)=VX(3,1)-DELT*(XXX(3,1)*RR(1,3) +XXX(3,2)*RR(2,3)
1+XXX(3,4)*RR(3,4)+XXX(3,5)*RR(3,5))
 VY(3,3)=VY(3,1)-DELT*(YYY(3,1)*RR(1,3) +YYY(3,2)*RR(2,3)
1+YYY(3,4)*RR(3,4)+YYY(3,5)*RR(3,5))
 VZ(3,3)=VZ(3,1)-DELT*(ZZZ(3,1)*RR(1,3) +ZZZ(3,2)*RR(2,3)
1+ZZZ(3,4)*RR(3,4)+ZZZ(3,5)*RR(3,5))
1-DELT*.98
 VX(4,3)=VX(4,1)-DELT*(XXX(4,1)*RR(1,4) +XXX(4,2)*RR(2,4)
1+XXX(4,3)*RR(3,4)+XXX(4,5)*RR(4,5))
 VY(4,3)=VY(4,1)-DELT*(YYY(4,1)*RR(1,4) +YYY(4,2)*RR(2,4)
1+YYY(4,3)*RR(3,4)+YYY(4,5)*RR(4,5))
 VZ(4,3)=VZ(4,1)-DELT*(ZZZ(4,1)*RR(1,4) +ZZZ(4,2)*RR(2,4)
1+ZZZ(4,3)*RR(3,4)+ZZZ(4,5)*RR(4,5))
1-DELT*.98
 VX(5,3)=VX(5,1)-DELT*(XXX(5,2)*RR(2,5)
1+XXX(5,3)*RR(3,5)+XXX(5,4)*RR(4,5))
 VY(5,3)=VY(5,1)-DELT*(YYY(5,2)*RR(2,5)
1+YYY(5,3)*RR(3,5)+YYY(5,4)*RR(5,4))
 VZ(5,3)=VZ(5,1)-DELT*(ZZZ(5,2)*RR(2,5)
1+ZZZ(5,3)*RR(3,5))+ZZZ(5,4)*RR(4,5))
1-DELT*.98
 KK=KK+1
 IF (KK.GT.50000) GO TO 83
 DO 9875 I=1,5
 XX(I)=ABS(X(I,3)-X(I,2))
 YY(I)=ABS(Y(I,3)-Y(I,2))
 ZZ(I)=ABS(Z(I,3)-Z(I,2))
 VXX(I)=ABS(VX(I,3)-VX(I,2))
 VYY(I)=ABS(VY(I,3)-VY(I,2))
 VZZ(I)=ABS(VZ(I,3)-VZ(I,2))
9875    CONTINUE
 INDEX=-1
 DO 600 I=1,5
 IF (XX(I).GT.0.000000000005) INDEX=+1
 IF (YY(I).GT.0.000000000005) INDEX=+1
 IF (ZZ(I).GT.0.000000000005) INDEX=+1
 IF (VXX(I).GT.0.000000000005) INDEX=+1
 IF (VYY(I).GT.0.000000000005) INDEX=+1
 IF (VZZ(I).GT.0.000000000005) INDEX=+1
600    CONTINUE
 IF (INDEX.EQ.+1) GO TO 105
 DO 1008 JJ=1,5
 X(JJ,2)=X(JJ,3)
 Y(JJ,2)=Y(JJ,3)
 Z(JJ,2)=Z(JJ,3)
```

```
      VX(JJ,2)=VX(JJ,3)
      VY(JJ,2)=VY(JJ,3)
      VZ(JJ,2)=VZ(JJ,3)
1008  CONTINUE
C WE NOW DETERMINE DATA RELATIVE TO P1.
      X(2,2)=X(2,2)-X(1,2)
      Y(2,2)=Y(2,2)-Y(1,2)
      Z(2,2)=Z(2,2)-Z(1,2)
      X(3,2)=X(3,2)-X(1,2)
      Y(3,2)=Y(3,2)-Y(1,2)
      Z(3,2)=Z(3,2)-Z(1,2)
      X(4,2)=X(4,2)-X(1,2)
      Y(4,2)=Y(4,2)-Y(1,2)
      Z(4,2)=Z(4,2)-Z(1,2)
      X(5,2)=X(5,2)-X(1,2)
      Y(5,2)=Y(5,2)-Y(1,2)
      Z(5,2)=Z(5,2)-Z(1,2)
      X(1,2)=0.0
      Y(1,2)=0.0
      Z(1,2)=0.
      VX(2,2)=VX(2,2)-VX(1,2)
      VY(2,2)=VY(2,2)-VY(1,2)
      VZ(2,2)=VZ(2,2)-VZ(1,2)
      VX(3,2)=VX(3,2)-VX(1,2)
      VY(3,2)=VY(3,2)-VY(1,2)
      VZ(3,2)=VZ(3,2)-VZ(1,2)
      VX(4,2)=VX(4,2)-VX(1,2)
      VY(4,2)=VY(4,2)-VY(1,2)
      VZ(4,2)=VZ(4,2)-VZ(1,2)
      VX(5,2)=VX(5,2)-VX(1,2)
      VY(5,2)=VY(5,2)-VY(1,2)
      VZ(5,2)=VZ(5,2)-VZ(1,2)
      VX(1,2)=0.0
      VY(1,2)=0.0
      VZ(1,2)=0.
      K=K+1
      IF (MOD(K,KPRINT).GT.0) GO TO 82
      V1SQ=VX(1,2)*VX(1,2)+VY(1,2)*VY(1,2)+VZ(1,2)*VZ(1,2)
      V2SQ=VX(2,2)*VX(2,2)+VY(2,2)*VY(2,2)+VZ(2,2)*VZ(2,2)
      V3SQ=VX(3,2)*VX(3,2)+VY(3,2)*VY(3,2)+VZ(3,2)*VZ(3,2)
      V4SQ=VX(4,2)*VX(4,2)+VY(4,2)*VY(4,2)+VZ(4,2)*VZ(4,2)
      V5SQ=VX(5,2)*VX(5,2)+VY(5,2)*VY(5,2)+VZ(5,2)*VZ(5,2)
      ENERGY=-A/((R(1,2)**3))-A/((R(1,3)**3))
     1-A/((R(1,4)**3))-A/((R(2,3)**3))
     1-A/((R(2,4)**3))-A/((R(2,5)**3))
```

```
       1-A/((R(3,4)**3))-A/((R(3,5)**3))
       1-A/((R(4,5)**3))
       1+A/((R(1,2)**5))+A/((R(1,3)**5))
       1+A/((R(1,4)**5))+A/((R(2,3)**5))
       1+A/((R(2,4)**5))+A/((R(2,5)**5))+A/((R(3,4)**5))
       1+A/((R(3,5)**5))+A/((R(4,5)**5))
       1+.98*(Z(2,2)+Z(3,2)+Z(4,2)+Z(5,2))
       ENERGY=ENERGY+(V1SQ+V2SQ+V3SQ+V4SQ+V5SQ)
       ENERGY=.000001*ENERGY
       WRITE (23,8000) K,ENERGY,RN(1,2),RN(1,3),RN(1,4),
       1RN(2,3),RN(2,4),RN(3,4),RN(2,5),RN(3,5),RN(4,5),XTIME
8000   FORMAT (I13,F12.6.9F7.3.F12.5)
       XBAR=(X(1,2)+X(2,2)+X(3,2)+X(4,2)+X(5,2))/5.
       YBAR=(Y(1,2)+Y(2,2)+X(3,2)+Y(4,2)+Y(5,2))/5.
       ZBAR=(Z(1,2)+Z(2,2)+Z(3,2)+Z(4,2)+Z(5,2))/5.
C NOTE THAT XBAR,YBAR,ZBAR ARE RELATIVE TO P1 SO THAT
C WE CAN EASILY PLOT THE MOTION OF THE MASS CENTER.
       WRITE (24,938) XBAR,YBAR,ZBAR
938    FORMAT (3F16.10)
82     IF (K.LE.10000000) GO TO 65
8383   WRITE (22,10) (X(I,2),Y(I,2),Z(I,2),VX(I,2),
       1VY(I,2),VZ(I,2),i=1,5)
       WRITE (22,8050) XTIME
8050   FORMAT (F15.5)
83     STOP
       END
```

Appendix A5
Newtonian Iteration Formulas

For a system of n equations in n unknowns, say,

$$f_1(x_1, x_2, x_3, \ldots, x_n) = 0$$

$$f_2(x_1, x_2, x_3, \ldots, x_n) = 0$$

$$f_3(x_1, x_2, x_3, \ldots, x_n) = 0$$

$$\vdots$$

$$f_n(x_1, x_2, x_3, \ldots, x_n) = 0$$

in which n can be large, a direct extension of the one dimensional Newtonian iteration formula yields the iteration formulas

$$x_1^{(k+1)} = x_1^{(k)} - \omega \frac{f_1(x_1^{(k)}, x_2^{(k)}, x_3^{(k)}, \ldots, x_n^{(k)})}{\partial f_1(x_1^{(k)}, x_2^{(k)}, x_3^{(k)}, \ldots, x_n^{(k)})/\partial x_1}$$

$$x_2^{(k+1)} = x_2^{(k)} - \omega \frac{f_2(x_1^{(k+1)}, x_2^{(k)}, x_3^{(k)}, \ldots, x_n^{(k)})}{\partial f_2(x_1^{(k+1)}, x_2^{(k)}, x_3^{(k)}, \ldots, x_n^{(k)})/\partial x_2}$$

$$x_3^{(k+1)} = x_3^{(k)} - \omega \frac{f_3(x_1^{(k+1)}, x_2^{(k+1)}, x_3^{(k)}, \ldots, x_n^{(k)})}{\partial f_3(x_1^{(k+1)}, x_2^{(k+1)}, x_3^{(k)}, \ldots, x_n^{(k)})/\partial x_3}$$

$$\vdots$$

$$x_n^{(k+1)} = x_n^{(k)} - \omega \frac{f_n(x_1^{(k+1)}, x_2^{(k+1)}, \ldots, x_{n-1}^{(k+1)}, x_n^{(k)})}{\partial f_n(x_1^{(k+1)}, x_2^{(k+1)}, \ldots, x_{n-1}^{(k+1)}, x_n^{(k)})/\partial x_n}$$

in which $0 < \omega < 2$. If the given system is linear, then the iteration formulas are called successive-overrelaxation (SOR).

A simple, often convenient modification of the iteration formulas uses $\omega = 1$, while the $(k + 1)$-iterates on the right-hand sides are replaced by k-iterates.

In general, matrix generalizations of Newtonian formulas have theoretical, but not practical, value when n is large.

Index

(8.7016124)

WBA

6700549

Nov 72

Sonn?

47 '13